INTRODUCTION
TO THE
PHILOSOPHY OF SCIENCE

Introduction to the Philosophy of Science

ANTHONY O'HEAR

CLARENDON PRESS · OXFORD

Oxford University Press, Walton Street, Oxford OX2 6DP

Oxford New York
Athens Auckland Bangkok Bombay
Calcutta Cape Town Dar es Salaam Delhi
Florence Hong Kong Istanbul Karachi
Kuala Lumpur Madras Madrid Melbourne
Mexico City Nairobi Paris Singapore
Taipei Tokyo Toronto

and associated companies in
Berlin Ibadan

Oxford is a trade mark of Oxford University Press

Published in the United States by
Oxford University Press Inc., New York

British Library Cataloguing in Publication Data
O'Hear, Anthony
Introduction to the philosophy of science.
1. Science. Philosophical perspectives
I. Title
501
ISBN 0-19-824814-8
ISBN 0-19-824813-X (Pbk)

Library of Congress Cataloging in Publication Data
O'Hear, Anthony.
An introduction to the philosophy of science.
Bibliography: p. Includes index.
1. Science—Philosophy. 2. Science—Methodology.
I. Title.
Q175.0454 1989 501 88-28884
ISBN 0-19-824814-8
ISBN 0-19-824813-X (Pbk.)

7 9 10 8 6

Printed in Great Britain by
Richard Clay Ltd
Bungay, Suffolk

For Jacob

Foreword

An introduction to a topic hardly needs an introduction itself. My aim in this book has been simply to introduce students and others to the philosophy of science, and to do so in a balanced way. That is to say, I have tried to lay out some of the central philosophical problems raised by natural science so as to show what can be said on various sides of a given issue. In the first chapter I have indicated why I think the philosophy of science is important and what I take its scope to be.

I should like to thank Angela Blackburn of Oxford University Press for encouraging me to write this book and for her help with the writing. Intellectually I owe a considerable debt to a number of friends and colleagues who were kind enough to comment on what I had written: to Michael Redhead and Roger Scruton who commented on the whole manuscript, and to Donald Gillies and Graham Macdonald who went through particular chapters with me. The comments of an anonymous referee for Oxford University Press were also most helpful.

Finally, I am very pleased to be able once more to thank Beverley Toulson for typing and preparing a manuscript for me. Her help and efficiency have made my task much easier than it would otherwise have been.

Contents

I

Science as an Intellectual Activity

There is no institution in the modern world more prestigious than science. Nor is there an institution which, as a whole, is less controversial. It is true that there are those who object to some aspects of the contemporary applications of science, to the use of nuclear power, say, or to the side-effects of certain industrial products such as the private motor car. But those who protest about such things are usually quite happy to have their messages transmitted by the latest audio-visual technology and their persons conveyed by high-speed train or plane. And in a thousand and one other ways, their lives are unthinkingly dependent on devices which have been made possible only in the last two or three hundred years and only through scientific discoveries. There can hardly be more than a handful of people all over the world who would actually choose to live completely without electricity or antibiotics or synthetic fibres or plastics or radio or mechanical transport or electronics. In this sense, then, science and its discoveries are deeply uncontroversial: at a practical level they form the unquestioned horizons within which the vast majority of mankind live or would like to live. Objections to science and scientific research tend to be partial, to *some* aspects of the application of scientific knowledge, leaving unquestioned most of its applications. They also tend to be (in the bad sense) theoretical, affecting the way people talk rather than the way they actually live.

As well as informing the way we live, the discoveries of science cut across political and religious divisions to a considerable extent. Again this is partly to do with the effectiveness in application of scientific discoveries. Most citizens of most states want the material benefits scientific discoveries make possible, even when they want other things as well. Rulers and political leaders are by now thoroughly intimidated by hearing about Hitler's failure to get the atomic bomb owing to his objections to 'Jewish' science (which was supposed to underlie nuclear physics) and about the disastrous effects on Russian agriculture of the ideologically orthodox but biologically incorrect theories of T. D. Lysenko. And, of course, they are right to be so intimidated: science does cut through political ideology, because its theories are about nature, and made true or false by a nonpartisan nature, whatever the race or beliefs of their inventor, and however they conform or fail to conform to political or religious opinion. In a world in which technological success is crucial to any regime, no sane leader is going to jeopardize his or her chances by interfering with scientific research or its applications on ideological grounds.

As I will suggest in the final chapter of this book, not everything one finds in writings critical of 'science' or 'the scientific mentality' is completely misguided. There are certainly areas of human life—the most important areas, in fact—about which science as such can have nothing to tell us, and where the application of methods analogous to those of science can only be harmful. But because of the importance of science and of these questions it is important to be balanced and honest in what one says about science, and to recognize both our dependence on it and its very real intellectual and moral merits.

On our dependence on science, it is simply not possible for the present population of the world to be supported at all, let alone enjoy a comfortable standard of existence with a reasonable life expectancy, without reliance on many of the discoveries of modern science. It should be obvious that there are not the space or resources in the world for a general return to nature. Just how we use technology, though, and just which technologies we attempt to develop are in a broad sense political questions, which require ethical and political decisions, and may be hotly debated. But debating such issues does not entail a globally anti-scientific stance, and those opposed to aspects of the nuclear industry, say, or to genetic engineering do their cause no good at all by adopting such a stance. Science is too prestigious to make global anti-science seem reasonable or politically attractive. The prestige of science is not mere propaganda, but derives in part from the solid realization people pre-theoretically have of its benefits and of its pervasiveness in the lives of us all. On a more theoretical level, though, its prestige stems from the way knowledge grows in science in contrast to what happens to other fields of knowledge, and from what is taken to be the objectivity of its claims.

In this book, we shall be mainly concerned to see how far science deserves its prestige on this theoretical level; we will attempt to see how far it can genuinely claim to present us with more and more true knowledge about the nature of the world. And for this, we shall have to concentrate mainly on the theories of science, rather than on its technological applications. For if true knowledge is growing in science, this means that the theories of science must be giving us more and more truths about the world.

Growth of Knowledge

In a perfectly obvious sense, over the last four hundred years or so there has been progress in science. Measurement of physical quantities becomes more precise, previously unknown particles and substances are discovered, new effects are produced and applied. Even if the ancients were wiser than us, and knew better how to live, they did not know the speed of light or the mass of the earth or the structure of the hydrogen atom or how to produce and apply lasers or the photo-electric effect. And we know now that, despite the brilliance of many of the observations, most of the theories of the universe expounded in Aristotle's writings are quite simply false. The earth is not the centre of the universe, nor do heavy bodies seek the centre of the earth as their natural resting-place, nor is the earth surrounded by series of concentric circles on which the other heavenly bodies revolve.

There is a striking contrast here between the development of modern science and the arts. No one would say of a work of music or literature that it was better than an earlier work just because it was later. A recent work by Luciano Berio is not, because of its modernity, better than Beethoven's Violin Concerto, and, it would be just as hard to say that Brahms's Violin Concerto was better than Beethoven's. In contrast to the development of theories in modern science, a later masterpiece in a given artistic genre is not thereby better than an earlier one, nor does it necessarily attain the aim of the genre better, or anything of that sort. Indeed, in the case of violin concertos, it would be hard if not actually senseless to specify a target at which all violin concertos are aiming, and in relation to which one could say that concerto A got nearer to the target than concerto B. Leaving Berio aside, it hardly makes sense to say of the Beethoven and the Brahms that one is better than the other. There just is no scale on which one could judge such things, once a certain level of expressive adequacy has been reached.

The case with scientific theories, though, is quite different. Here we are able to specify a clear target at which all theories aim, and we often have confidence that theory A has got closer to the target than theory B. The aim might be characterized as discovering the truth about the natural world, and when we have theories which aim to describe the same bits of the natural world we can often say that a later theory is better than an earlier. Thus, Copernicus's heliocentric picture of the universe was better than Aristotle's geocentric picture, and Newton provided a better account of the solar system and the universe than either. Similarly, every schoolchild now knows that the chemist Joseph Priestley mistakenly thought he had identified a substance he called 'phlogiston', a gaseous fluid with possibly negative weight which, among many other things, was given off when me als burned. We now know that there is no such thing and that there is no one substance that does all the things phlogiston had been supposed to do. Priestley had actually been observing the effects of oxygen, which metals absorb from the air when burned, rather than those of the phlogiston they were supposed to give off. When he imagined he had isolated a sample of dephlogisticated air, he had actually produced oxygen.

As I say, this story about Priestley and phlogiston is something every schoolchild knows and in that sense, every schoolchild has knowledge Priestley did not have, and in that sense is in advance of Priestley. But it is not the case that every schoolchild is a better chemist than Priestley, any more than Berio or even Brahms is a better composer than Beethoven. We can speak of an objective advance or progress in the one case but not in the other because we can speak unambiguously of knowledge having grown in the sciences, so that those working later in a given field of science, by that very fact alone, may be said to know more than their predecessors.

Apparently rather against the grain of what I have just been saying, T. S. Eliot once remarked that in literature we know far more than our predecessors because what we know is their work. Growth of knowledge in science, though, is not at all like that, and this is another way of bringing out the distinction I am drawing between science and the arts. Although a few stories from the history of science, like that of Priestley and phlogiston, are part of the folklore of the subject, most workers in a scientific field do not know the history of their field in any depth or detail. They do not have to know it, because the history of science will consist largely of theories that have been discarded, and which are regarded as giving far less true information about the world than their successors. An astronomer will tell you what, in the considered view of astronomers today, the universe is like, not what Aristotle or Copernicus thought it was like. The theories of Aristotle and Copernicus are, from the scientific point of view, dead; we have progressed beyond them and there is no need to revive them except as historical curiosities we may briefly contrast with present knowledge. The case is quite different with works of art and literature. The dead writers of whom Eliot spoke are part of the soil and tradition in which we live, and we deepen and refresh our understanding both of ourselves and of art by returning to them and deepening our acquaintance with them.

Objectivity and the External World

The reason why we in doing science have no need to return to past science is because the theories of science are not about human endeavour or human expressiveness. Human self-expression and understanding is a cumulative, historical process in which where we are now and what we now think of ourselves is rooted in the forms of life and expression developed in the past, and will always involve some coming to terms with our history and our past. But a scientific theory will, by contrast, be dealing with a world independent of human history and human intervention. The truths science attempts to reveal about atoms and the solar system and even about microbes and bacteria would still be true even if human beings had never existed. As we have noted, it is a humanly impartial ahistorical nature that decrees the truth or falsity of scientific theories, and it does so without regard to religious or political rectitude.

This brings us to one of the distinctive features of scientific activity, which morally and humanly is one of its great strengths. The impartiality of nature to our feelings, beliefs, and desires means that the work of testing and developing scientific theories is insensitive to the ideological background of individual scientists. A scientific theory will characteristically attempt to explain some natural phenomena by producing some general formula or theory covering all the phenomena of that particular type. From this general formula, it will be possible to predict how future phenomena in the class in question will turn out. Whether they do or not will depend on nature rather than on men, and any scientist can observe whether they do or not, regardless of his other beliefs.

To take a concrete example, Newton produced a set of formulae which give us a general account of the motions of bodies, showing how these motions are affected by such things as force, mass, acceleration, and gravitational attraction. Among other things, these formulae explained the courses the various planets took in orbit around the sun. From Newton's formulae, it was possible to predict the future behaviour of the planets that were already known, and, as it turned out, the very existence of planets unknown in Newton's time. We now know that Newton himself had all sorts of theological and mystical concerns, and that these may well have inspired his search for general mathematical formulae uniting events on earth and in the heavens. But his theories were intelligible to people who did not share these concerns, and his predictions were testable by anyone who knew how to make the appropriate observations, regardless of their ideology, race, or upbringing. You do not have to share Newton's outlook in any way, in order to come to a reasoned assessment of the truth or otherwise of his theories, for the observations relevant to such assessments are of a world not created by us, the perception of which does not crucially depend on one's ideological standpoint. The case is quite otherwise with some of the grand theories of psychology and the social sciences, where critics are sometimes told that their criticisms are invalid because their observations are distorted by their being sexually repressed (as in the case of Freudianism) or because they are not identifying themselves with the proletariat (as in the case of Marxism). But, because of the nature of the enterprise, the scientific community is non-sectarian and its work cuts across all sorts of human divisions. There is no such thing as British science, or Catholic science, or Communist science, though there are Britons, Catholics, and Communists who are scientists, and who should, as scientists, be able to communicate fully with each other. The ideological or religious background of a scientist becomes important only when, as with a doctrinaire Marxist-Leninist like Lysenko or some fundamentalist Christians, non-scientific beliefs make disinterested scientific enquiry impossible.

Prediction and Explanation

In the previous section it was asserted that the theories of science characteristically take the form of general mathematical formulae covering a particular range of types of event, from which it is possible to deduce predictions of specific events. Newton's laws, for example, give us general formulae concerning the motions and mutual attraction and repulsion of heavy bodies, from which we can predict such things as solar eclipses. From the standpoint of modern science, there is a close connection between the notions of prediction and explanation. If you can produce general formulae allowing you to make mathematically precise predictions of a class of specific states of affairs, you will generally have gone a good way to providing an explanation of those states of affairs. To take another example, the classical gas law tells us that the volume of any body of gas is a function of its temperature and pressure ($V = c.T/P$, where c is a constant factor). Applying this formula to specific bodies of gas, with particular temperatures and pressures which we measure, enables us to predict their actual volumes; in thus reaching a prediction of a specific instance on the basis of a general formula covering all instances of a given type, it is quite natural to think we have explained the relevant characteristics of the specific instance.

One reason for not saying here that we have *always* gone some way to producing an explanation when we are able to make predictions on the basis of general formulae is that there are cases discussed in the philosophical literature in which one is enabled to produce a precise prediction of states of affairs on the basis of a general theory without—it is alleged—being tempted to say that one has any sort of explanation before one. Thus, for example, by invoking Pythagoras's theorem, one can predict the distance of a mouse from an owl, when all we knew was that the mouse was four feet from a three-foot flag-pole on top of which was an owl; but, it is said, one would not want to say that the theorem explained the distance of the mouse from the owl. Against this example it might be said that there was no genuine prediction here, in the sense of an inference from a past state of affairs to a future one, as opposed to a move from a state of past ignorance to one of future knowledge. It is not clear, though, that all scientific explanations do involve predictions from past states of affairs to future ones, rather than predictions about what one will find on the basis of existing knowledge, for this latter type of reasoning is involved when people deduce conclusions concerning the nature of the big bang from their cosmological theories and their knowledge of the current state of the universe. The predictions by which one tests such speculation may well be predictions about what one will find when one probes traces of past events. However, given that we are prepared to work with a concept of prediction which is wide enough to encompass the prediction and discovery of as yet unknown facts, including facts about the past, it is certainly the case that we now expect scientific explanations to have predictive power. We can say this even though there may be cases, like that involving the Pythagoras theorem, when we can make predictions, or at least deduce as yet unknown facts, on the basis of general theories, without wanting to speak of an explanation of those facts. The reason why many criticize Freudians and Marxists for being unscientific is precisely because their theories either lead to no specific predictions at all or to predictions that are false. Making predictions on the basis of one's theories is, then, a necessary if not sufficient condition for a genuine scientific explanation.

The notion of a scientific explanation was not always linked so closely to its mathematical and predictive power. In the science associated with Aristotle and his followers, giving an explanation of a phenomenon consisted in delineating its essence, or essential properties, and in showing why, in order to fulfil its function or nature, it had to have those properties. Fire rose, for example, in order that it should reach its natural resting-place, which was taken to be a spherical shell just inside the orbit of the moon. The essence of fire, being a light body, was to rise. It does so in order to fulfil its nature.

From the modern scientific viewpoint there are at least two things wrong with this 'essentialist' type of explanation. In the first place, we have no justification for imputing purposes to natural phenomena like fire or planets or heavy bodies. Their activity is conditioned by the forces that act upon them, their underlying structure, and the interaction of the two. They do not have any ulterior purposes, or essential nature they are trying to fulfil. Secondly, there is nothing in a typical Aristotelian explanation about precise quantities or measurements. They give us reasons (of a sort) for why things happen, but not the precise amounts or distances or times involved. And these precise measurements are crucial for modern science, because they are required for the formulation and application of its theories.

It is easy to see why the shift occurred from Aristotelian essentialist explanations to the mathematical-predictive explanations of modern science. If you want to control and manipulate phenomena, then what you need to know are the precise conditions in which effects of given sort occur. If you are working with a piece of metal, you want to know just how much it will expand under given degrees of heat. You do not want to be told that its expansion is due to the fact that it has to expand in order to fulfil its nature. And, as we shall see in the next chapter, modern science is very much about controlling nature, hence its tendency to elide prediction and explanation, and the reason why its predictions will characteristically, if not universally, be predictions about states of affairs which have not yet happened.

Yet, even at this point, one might feel that there is something to be said for a more meaty type of explanation than appears to be given in simply producing formulae for prediction. Newton himself gave expression to this feeling when, at the end of his *Mathematical Principles*, he said that while he had demonstrated the reality of gravity and its effects—by precise mathematical methods, we would stress—he had not yet been able to explain the cause of these effects. It is as if a purely mathematical correlation of events, saying, for example, that the gravitational force on such and such an object will be so and so in such and such circumstances, stays too much on the surface of things, and fails to give us insight into the underlying structure of gravitational phenomena or of the essence of gravity. We shall take this point up again in Chapters 3, 5, and 6, when we consider whether a full scientific explanation is *more* than a device for predicting effects in the natural world. But we have just seen a good reason for thinking it must be at least that.

2

Induction

Baconian Induction

How do scientists arrive at their theories and explanations?
In the last chapter, we touched briefly on the Aristotelian
method, which depended on some sort of intuition of the
essential properties and natural purposes of things. But by
the seventeenth century, Aristotelian science had become
thoroughly discredited in the light of new discoveries in
astronomy, anatomy, physics, and elsewhere, and its methods were generally blamed for having held up scientific
progress for centuries. There was, moreover, a new emphasis
on controlling and manipulating nature, 'for the relief of
man's estate', as Francis Bacon put it, against the Aristotelian ideal of a disinterested philosophical contemplation of
the world and its order and harmony. Bacon is, in fact, often
and not altogether incorrectly taken to be the spokesman for
the new spirit and the new science.

Bacon's ideas about scientific method have subsequently
become known as Baconian induction. This method is
expounded in his *Novum Organum* (published in 1620) and
in fact still forms the basis of what many people think of as
the method of science. Indeed, the notion of science as a
progressive accumulation of knowledge about the material
world, which we considered in Chapter 1, was apparently
first propounded by Bacon.[1] And he believed that he had hit

---

[1] On this point, see Anthony Quinton's *Francis Bacon* (Oxford University
Press, 1980), p. 30. In the following paragraphs, I have drawn on Quinton's
excellent summaries of Bacon's thought.

on a method by which this accumulation would become much more likely than if his precepts were neglected.

In the *Novum Organum*, there are both negative and positive doctrines. The negative doctrine is picturesquely expounded by Bacon in terms of four types of 'idols' which have dominated and distorted men's minds, delaying the acquisition of true knowledge. The 'idols of the tribe' are tendencies we all have to see things in relation to us rather than as they are in themselves. In Bacon's view man is definitely not the measure of all things and we unthinkingly tend to impose order on phenomena which is not there. If we would command nature, we must first learn to obey her. Then there are the 'idols of the cave', which are the predispositions of character and learning with which different individuals approach the facts, rather than seeing them as they really are. The 'idols of the market' arise through the use of language, when we read back into nature conceptions which have arisen simply through our using words which actually stand for nothing (such as 'Fortune, the Prime Mover, Planetary Orbits, the Element of Fire, and like fictions').[2] Finally, there are the 'idols of the theatre', which are due to the malign influence of philosophical systems on our minds. In Bacon's view we should not be misled by Aristotle's talk of experimentation and observation. Someone in the grip of a philosophical system, including Aristotle himself,

had come to his conclusion before [he did his experiments]; he did not consult experience, as he should have done, for the purpose of framing his decisions and axioms, but having first determined the question according to his will, he then resorts to experience, and bending her into conformity with his placets, leads her about like a captive in a procession.[3]

The effect of Bacon's negative doctrine is that any properly established science will have to begin from and be controlled by observations untainted by the presuppositions

[2] Bacon, *Novum Organum*, i. lx.　　[3] Bacon, *Novum Organum*, i. lxiii.

of the idols, or any other sort. We have to approach nature with an innocent and uncorrupted eye, and preserve this innocence through our researches. For Bacon, the true scientist will be the paradigm of the objective observer who frees men from the illusions and myths of the past.

The presuppositionless observation required by Bacon is not, however, conducted in a random or disorganized way. The scientist is not a spider, spinning webs out of his own fancyings, but neither is he an ant who only collects; he is rather a bee, gathering materials from 'the flowers of the garden and of the field', but transforming and digesting it 'by a power of its own'.[4] The scientist-bee will proceed by preparing 'a natural and experimental history' of all the things relevant to the object or phenomenon he is interested in investigating, and he will then tabulate the material he has so gathered in order to analyse it and so discover the true nature of what he is interested in, and understand the conditions which produce the object or phenomenon in question.

Bacon's own example of a scientific investigation is that of discovering the nature of heat. We are to begin by collecting all known instances of heat and describing the features present in each, even though the instances are quite different from each other. Indeed, it is important for the success of the task that they should be, and that we make our collection 'without premature speculation, or any great amount of subtlety'.[5] Having gathered and described examples of heat, ranging from the rays of the sun, through bodies rubbed violently, boiling liquid, compost and horse dung, to the effects on us of fortified spirits of wine and aromatic herbs, we then gather and describe cases which are similar in some significant way to the heat cases, but lacking heat itself. So we note down the rays of the moon and stars, non-boiling liquids, ignis fatuus or will-o'-the-wisp, air in an unheated

---

[4] *Novum Organum*, I. xcv.   [5] *Novum Organum*, II. xi.

state, and so on, trying all the time to see if there are cold
examples of the things which in certain conditions are hot.
We also draw up lists of different degrees of heat, so as to see
when heat is more or less present. Then, by comparing the
various lists and the features present and absent in the
various cases, we may ascertain what the essential nature of
heat is, and what is merely an incidental accompaniment to it
in specific conditions. The essential nature is, of course,
constituted by the features which are present in all the
positive instances, absent in all the negative instances (where
heat is not produced), and vary in appropriate degrees in the
different cases. The process of excluding the incidental
accompaniments is for Bacon the crucial activity of the
scientist, for in this way we come to see that only some
features are present in every positive instance, and absent in
every negative instance. Bacon concludes that heat is a
'motion' which acts on the 'smaller particles' of bodies
(which is why there was no negative instance to the case of
bodies being rubbed violently), and so you will undoubtedly
produce heat if you 'excite a dilating or expanding motion' in
any natural body.[6]

What Bacon is proposing is a set method by which
scientists can produce theories which, according to him, will
have a better chance of being true than theories produced by
other methods. He thinks that his stress on negative instances helps to overcome some of the difficulties involved in
basing a theory simply on positive evidence. The natural
tendency of the human mind is rashly to generalize on the
basis of all too slender samples of cases and to find order and
generality where there is none. Bacon's search for negative
instances is designed to rule out such rash generalization.
From observation of the sun and of fires, we might initially
think all heat is connected to rays of some sort, but we test
this hypothesis, and eliminate it when we find lunar rays,

---

[6] *Novum Organum*, II. xx.

which are cold. He also places considerable stress on cases
where some observation provides a testing ground between
two competing theories. Bacon, in other words, prefers as
evidence what amounts to the survival by theories of genuine
tests to the simple piling up of evidence in favour of theories,
and in this he anticipates the work of Sir Karl Popper, as we
shall see.

But what of his claim to have produced a better method for
science than that of his predecessors? Our answer to this
question will fall into two parts, and will deal with what
might, at first sight, seem the most attractive and 'scientific'
features of Bacon's methodology. These are the belief that
science ought to proceed by means of presuppositionless
observation and the idea that scientific research can be
conducted by means of the systematic tabulation of data.
Both these proposals assume that the proper method for
science is the elimination of human subjectivity and guesswork, and both suffer from the fact that such an elimination
is neither possible nor desirable in scientific research.

Presuppositionless Observation

Bacon's philosophy of science seems attractive because it
recommends a thorough cleaning of our mental slate. It came
at a time when there was optimism about science and its
possibilities, and when the effects of centuries of obscurantism seemed about to be swept away. And how better to sweep
them away than by cleansing the mind of all its presuppositions and prejudices, and reading the book of nature with
fresh eyes?

Unfortunately for the usefulness of Bacon's ideas and also
for the possibility of any project which appears to require a
presuppositionless reading of the book of nature, we cannot
make any observations without some ideas concerning the
nature of what it is we are observing. All our observations are

conditioned by a sense of what type of thing or property in our environment is to be focused on. This sense may be pre-theoretical, as when we 'naturally' notice bright colours or moving animals or what J. L. Austin memorably referred to as the 'medium-sized dry goods' in our vicinity, or it may be more theoretically inspired when, as with Bacon's own example, we go round the world actively looking for examples of heat. In either type of case, the idols of our mind are stimulating us to pick out some features of our environment, to the exclusion of the infinite variety of other features we could have focused on had our sensory apparatus or interests been different.

This point becomes even more evident when we consider what is central to Bacon's tabular methodology, the picking out of repetitions of types of case, in Bacon's own example, that of heat phenomena. It is, of course, an assumption on Bacon's part that all the instances he is taking as examples of heat are actually examples of the same natural kind, a natural kind being a group of phenomena occurring naturally in the physical world with the same underlying physical constitution. Thus water is said to form a natural kind, by virtue of its constitution as H_2O, and so are members of biological species, by virtue of their shared genetic structure. On the other hand, things we group together in ordinary discourse under some category may not constitute a natural kind in this sense. It is very doubtful that all Bacon's instances of heat have a common underlying nature, and actually are all members of the same natural kind.

In fact, any two objects or events can, under some category or other, be seen as similar to another. A man stepping on a bus, a child throwing a ball, and a secretary typing away at a desk may not seem very similar on the face of it, but they can all be seen as examples of human activity, and perhaps should be so seen if we were drawing up Baconian lists of actions initiated by human agents. Similarly, an apple falls to the ground, the moon influences a high tide, and the earth goes round the sun. These are, for Newton, all examples of gravity in operation, but he needed a remarkable theory to see this and one which went against a centuries-old assumption that heavenly phenomena and earthly ones were subject to quite different types of law. Here, having an idol in his mind proved scientifically highly significant. Would the supposedly empty mind of a Baconian scientist ever have listed such apparently different things together? In this case, far from a mental presupposition impeding scientific progress, progress actually depended on it. On the other hand, the things which one naturally lumps together (though not, to be sure, presuppositionlessly) may not from the causal point of view be repetitions at all. In Bacon's own case, heat in boiling water and heat in distilled spirits surely have quite different natures and causes, and, scientifically speaking, are hardly repetitions of the same phenomenon or members of the same natural kind.

It is not simply that theories or presuppositions can help us to see scientifically relevant repetitions where we might not otherwise have done so or lead us to group dissimilars together. It is rather that without some presupposition, tacit or explicit, we will not see two events or objects as similar at all. For if there are respects in which any two events are similar, there will also be respects in which they will be dissimilar. A farmer may plough this field on two separate days, and for him the tasks will be thoroughly repetitious. But for the mouse whose nest is destroyed on the second day, the dissimilarity will be all important.

My point against Bacon is simply that in his drawing up his lists of instances he will have to classify the events in question under certain concepts, in order that the events can be seen as like or unlike each other, and in doing this he will necessarily be relying on presuppositions, picking out certain similarities and dissimilarities between the events as important, and ignoring others which seem unimportant for his purposes. Without assumptions of this sort, there will be no lists, and the method will not get going at all.

Bacon's model of presuppositionless observation is intended to elicit the true nature or cause of a particular type of phenomenon. The idea is that by listing the various features of, say, heat, we will find which are constantly conjoined with it, and so come to isolate the true cause. But this process is not going to be nearly as straightforward as Bacon appears to think. As Quinton puts it, 'he fails to see that causes may be spatio-temporally remote from their effects';[7] once this is realized, though, the very notion of the event or instance to be observed becomes hopelessly indeterminate. We have already pointed out that the movement of the tide is affected by the moon. Is the presence (or even the absence) of the moon part of an instance of tidal movement? Are we supposed to record *everything* surrounding the event we are interested in, in the hope that our tables will eventually reveal the other features of the environment which are always present when the type of thing we are interested in occurs? But this would be a hopeless and a futile task, for again without some idea of the relevant features of the surrounding environment we would never be able to complete any list of the accompaniments to an event. And there would in practice be precious little hope of bringing the position of the moon into our observations of tides without some pretty strong suspicion or presupposition of its relevance; its relevance might never strike us if all we did was to list the pre-theoretically obvious accompaniments to the tides.

An interesting example[8] of the sort of problem which would arise in the discovery of the true cause of an effect, if one were to follow strict Baconian principles, is given by the

[7] Quinton, *Francis Bacon*, p. 62.
[8] Due to Carl Hempel, *Philosophy of Natural Science* (Prentice-Hall, Englewood Cliffs, NJ, 1966), pp. 3–6.

story of Ignaz Semmelweis's discovery of the cause of the
puerperal fever which was plaguing the unfortunate women
who gave birth in one particular ward in the Vienna General
Hospital from 1844 to 1848. After testing and rejecting
numerous hypotheses as to the cause of this fever, which
related directly to the conditions of the women and their
treatment, and to comparisons with women in another
labour ward where there was no significant incidence of the
fever, Semmelweiss was alerted to the possibility of the true
cause by the death of a colleague from a rather similar type of
illness after cutting himself with a scalpel which had just
been used for an autopsy. It turned out, on further investigation, that doctors who attended the women in the ward in
question had often come straight from conducting autopsies,
without any proper disinfecting of their hands. Once this
practice was instituted, the childbed fever deaths dropped to
a tenth of what they had been, and a major landmark in the
discovery of the importance of micro-organisms in illness
had been passed. The point to draw from this example is that
no amount of systematic and presuppositionless observation
of the circumstances surrounding the puerperal fever cases
could have elicited the true cause. It needed the inspired
guess of Semmelweis that something quite outside the
immediate circumstances of the women was responsible for
their deaths.

The Baconian demand to observe in a presuppositionless
way in order to elicit the true causes of things presumably
entails that one notes down everything in the environment of
one's positive instances, or one would be using one's presuppositions in fixing on some things in the environment rather
than others. But one cannot simply observe *everything*, even
in a delimited segment of space and time. One needs some
sense of what one is to look at, and which features of the
environment are to be noted. And there is, in any case, no
guarantee that the true cause of the effect one is investigating
is going to be in the particular segment of space and time one

fixes on as constituting the relevant environment of the effect.

Observation without presuppositions or idols, then, is not possible. In observing anything we are inevitably reacting to our environment in the light of dispositions to notice some things rather than others. These dispositions can be quite pre-theoretical, built into our sensory apparatus or our first language. Normally sighted human beings are physiologically disposed to notice the colours of things, and also to group certain different colours as similar to each other and dissimilar from other colours. We group a pale blue and a dark blue as more similar to each other than either is to green, for example. And our language can lead us to make finer and more precise colour classifications. But there are creatures which do not react to colour at all and, more significantly, Newtonian and post-Newtonian science tells us that colour is not a fundamental property of the world. So to group things in terms of colours would be scientifically pointless. Science will teach us to look for quite other respects in which to group and classify objects.

If presuppositionless observation is not possible, then a large part of the Baconian methodology is undermined, and a lot of its apparent attractiveness goes. But it might still seem a good idea to stick as closely as possible in science to the patient collection and tabulation of data recommended by Bacon. By this aspect of Baconian induction one might hope to avoid the wilder flights of fancy and subjectivity.

The Role of the Imagination in Scientific Theorizing

Against this view of the scientific method, as essentially unimaginative, which is still widely regarded as ideal, we must point out that many of the most admired scientific theories were not produced in this sort of way, nor could they have been. Thus, for example, Kepler arrived at his three laws of planetary motion through a series of attempts to read mathematical correlations into the data. In asserting in his second law, that the radius vector of the sun to a planet sweeps over equal areas in equal times, he was going far beyond what the data available to him could give him grounds for asserting. For one thing, the law applies to all planets, including undiscovered ones, but even restricting his law to the then known planets in our solar system the observations actually available to him could not on their own have entitled him to assert anything as bold and simple as he did, with the planets all orbiting the sun with mathematical regularity. Tycho Brahe, whose observations Kepler had to a considerable extent used, had concluded from them that the moon and the sun orbited the earth, which was at the centre of the solar system. What Kepler was actually doing here, as elsewhere, was acting on the Pythagorean assumption that the world is organized on principles of mathematical simplicity and harmony, and attempting to read these harmonies in the simplest possible way into the data.

In considering Kepler's achievement, we have to remember that he was not simply reading mathematical conclusions from an already existing map of the solar system. To a large extent, he was drawing the map itself, as a response to his mathematical analysis of the data. Prior to Kepler, it had not been realized that planetary orbits were elliptical, nor could it be said that it was generally accepted that the sun was the centre of the solar system when even as acute an observer as Brahe denied it. Significantly, before formulating his laws of planetary motion, which had the planets orbiting the sun in ellipses, Kepler had attempted other mathematical analyses of planetary activity, based on the geometry of regular solids, and he subsequently calculated planetary densities as mathematically correlated with their distances from the sun, but these attempts proved to conflict too much with the observational data. In each case, successful and unsuccessful, Kepler, whilst basing his theories on the facts known to him at the time, was actually going far beyond them. In so doing, he produced bold and potentially highly testable theories, one set of which had the additional virtue of being close to the truth and was recognized to be such by Newton. But unless he had worked by deliberately going way beyond the actual available data, he would never have been inspired to produce his successful laws.

Kepler's laws of planetary motion were close enough to the truth to be regarded later as observational laws. That is to say, they seemed to do little more than cover the known observational data adequately, and to lead to successful predictions. From the Baconian point of view, it would seem that here was something which could stand as proven knowledge and would form the basis of future theories. And for this reason it is often claimed in histories of science that Newton's dynamics follows from the laws discovered by Kepler and Galileo, the idea being that Newton simply built on and integrated the results achieved by his predecessors.

But, as Popper has pointed out,[9] here too the Baconian model of the theorizing scientist working slowly and carefully upwards from the known observational data fails. For Newton did not simply incorporate Kepler's laws into his theories. His theories actually contradict and correct Kepler's laws. Kepler's third law provides us with a formula for giving the ratio between the orbital velocities of two planets and their distance from the sun. For Newton, though, this formula is at best an approximation, which both neglects the mutual attraction of the planets concerned and assumes, quite erroneously, that the masses of the planets are either equal to each other or negligible as compared to that of the sun. Even Kepler's first two laws, which describe the orbits of the planets as elliptical and the velocity of a planet in orbit are in fact only approximations to the actual orbits of the planets, for they do not take into account the interaction

[9] In his *Objective Knowledge* (Clarendon Press, Oxford, 1979), pp. 197–204.

of the planets with each other, and this too is made clear by Newton.

What emerges from our considerations of Kepler and Newton here is that there is much highly admired scientific work which does not proceed by means of the simple collection and tabulation of presuppositionless data. Even if we could regard observational data as relatively presuppositionless, neither Kepler nor Newton regarded themselves as obliged to stick as close as possible to the data they had. Kepler theorized boldly, even wildly, about the underlying map of the heavens, with only incomplete and highly partial data. Newton did this too, in his own way, and in doing so amended and improved existing 'observational' data.

Bacon was, of course, right to stress the way in which observation is central to modern science, but he was wrong to attempt to rule out speculation and presupposition altogether. No observations can be made at all, without some initial predispositions to notice some things rather than others. In systematically observing particular types of phenomena, more explicit presuppositions come into play about what it is we are observing. And in much scientific theorizing the mind goes far beyond the existing data, in search of an underlying cause or a unifying key to a whole range of phenomena. This may be so particularly when mathematical formulae applying to a whole range of phenomena are sought, and it is significant that Bacon himself was very weak on the importance of mathematics in modern science. Against the over-reactions of some of Bacon's later critics, it is fair to say both that random observations can play a part in scientific discovery and that a lot of day-to-day scientific work does consist of the systematic collection and analysis of data. But neither of these processes can be entirely presuppositionless, nor, if the history of science is anything to go by, can systematic fact-collecting and analysis be seen as the only desirable way to do science. If we live in a world where much is unknown to us, we are not likely to penetrate its

further secrets by piling up more and more information of
the sort we already have. To go further, we will need
precisely the inspired guess and the wild insight, where the
categories of thought which currently constrain our thoughts
and—*pace* Bacon—our observations are set aside. But it is
just this sort of setting aside of our current presuppositions
that Bacon's view of science would in effect deny us for it is
just when we think we are observing presuppositionlessly
that our observations will be full of unexamined received
ideas. Nor should we be afraid of the idols of the cave: the
fact that peop le have different presuppositions and temperaments
can help in setting aside preconceptions in a field of
knowledge. For this reason it is important that science
should be an open society in which people with different
presuppositions and backgrounds should be able to converge
on a common, but largely unknown, world.

Inductive Proof

So far in this chapter, we have been concerned with what is
generally known as the inductive method, and which was
initially associated with the name of Francis Bacon, and then
subsequently with John Stuart Mill. The inductive method
recommends, as we have seen, a stepwise ascent in science
from observation to theory. We begin by collecting the
relevant observations, as many as we can, and as far as
possible without presuppositions. We then tabulate the data,
so as to isolate the features which are constantly associated
with the phenomenon we are interested in, both positively, in
the sense of always being there when the phenomenon is, and
negatively, in the sense of never being there when the
phenomenon is not. If we find such features, we may then
infer that this is the cause of our phenomenon. In effect, at
this third stage, we will be saying that this cause will *always*
bring about that effect. We will be making a generalization

on the basis of our evidence, and we may then put the generalization to the test, by trying it out in various new conditions (fourth stage). Even if we are not after a strictly causal theory, but simply want to discover how various phenomena are correlated mathematically or in some other way, for the inductivist the procedure will follow the same four stages.

We have already seen a number of grounds for criticizing the inductivist account of the first two stages of enquiry. But what of the third and fourth stages? Leaving aside the precise route by which we reach our generalization in the third stage—whether the means is more Keplerian or more Baconian—how can we ascertain its truth? We will undoubtedly submit it to further tests (fourth stage), but how far can doing this establish its truth? The intuition of the inductivist here is that the combination of a large amount of evidence favourable to a general hypothesis with no evidence against it gives us good reason to think the hypothesis true. I shall call any such proposal, that a theory can gain increases in its probability of being true by receiving more and more supporting evidence, an inductive proof. As a terminological point, it is important to appreciate that accepting the validity of inductive proofs does not entail that you accept all (or any) of what Bacon thought about the first two stages of enquiry. We can, and should, separate the discussion of inductive method from inductive proof.

The obvious problem which arises in connection with an inductive proof of a scientific theory is that in such a proof the evidence for our theory always falls far short of being conclusive. As an example of a deductive argument, we can take the familiar syllogism:

All men are mortal.
Socrates is a man.
Therefore, Socrates is mortal.

As with all valid deductive arguments, the conclusion follows from the premisses. Socrates is mortal (conclusion) because all men are mortal and Socrates is a man (premisses). The premisses cannot be true and the conclusion false. The truth of the premisses guarantees the truth of the conclusion, which is the strength of a valid deductive argument, but also its weakness. The conclusion cannot go significantly beyond the premisses, even where it draws hitherto unknown consequences from the premisses. But the situation is quite other in the case of an inductive proof. Comparatively few observations of a few planets are supposed to support a theory claiming that all planets always behave in the specified way. Inductive arguments promise to extend our knowledge in a significant way, going way beyond our actual experience. The problem is to see how there could be any valid proof that so extends our knowledge. How can we be sure that cases we have not experienced will be like those we have?

The answer is that we cannot be sure. Obviously we would like to have such an assurance in order to plan our future on a firm basis but, as we learn a thousand times a day, the future differs significantly from the past in very many respects.

The uncertainty or invalidity of inductive proof has been well known to philosophers since the time of Hume, who argued that from the strict logical point of view we have no justification in generalizing from instances we have experience of to those of which we have no experience. Such generalizations are of the essence when it comes to scientific theorizing, where what we are trying to do is to draw conclusions about a whole class of events from the evidence of only a few. It is sometimes said that the difficulty with the theories of science is that they attempt to cover a potentially infinite range of cases, actual and possible, throughout the whole of space and time. Any evidence we have will necessarily be restricted to a finite number of cases, and will therefore be quite insignificant considered as a proportion of the total cases. Now while this certainly is a difficulty if we are concerned in the infinite case to express the ratio of

favourable cases to the total number of cases, it is not the
basic difficulty highlighted by Hume. Even if we knew that
there were only a hundred cases of a certain type of event,
and we had observed ninety-nine of them, how could we be
sure that the hundredth would be similar to the first ninety-
nine?

The answer is that without invoking an assumption to the
effect that the future will be like the past or that our first
ninety-nine cases were in some strong sense representative of
the total, we cannot be sure. This is the difference between
inductive and deductive argument, that inductive argument
extends our knowledge but only on certain rather strong
assumptions, which may, at the crucial moment, let us down.

Many people would be inclined to react to this difficulty
with inductive argument by saying that we are entitled to
rely on inductive proofs because we have often done so in the
past without coming to grief. But Hume had an answer to
this too. An argument of this sort as an attempt to justify
inductive argumentation would be quite hopeless. For in our
purported justification we are appealing to past evidence (of
the past success of induction), and it is precisely the validity
of such appeals to past evidence we are trying to find a
justification for.

Nor will it help to weaken inductive proof and say that
even if inductive proofs do not give us certainty, they
certainly make it *probable* that the next case will be like the
ones we have hitherto observed. For a Humean will ask why
past evidence should have any bearing on future likelihoods.
How do we know that the future is even *likely* to resemble the
past? And if we are concerned to base our assessment of the
probability of a general theory about the whole of nature on
finite evidence, then the point about the ratio of any finite
number of actually observed cases to a potential infinity of
further cases will hit us with a vengeance. For the probability
of a universal theory, like one of Newton's laws which refer
to all movements anywhere in the universe, will always be

zero, or at best vanishingly small, however much evidence we collect in its favour, unless we covertly rely on the assumption that the evidence we have is a good sample of the rest of the cases. But it is just this assumption that Hume's argument questions.

We have so far been considering the problem of induction in terms of the difficulty of knowing that future or unexperienced cases will be like those we have so far experienced. Actually, though, as Nelson Goodman has pointed out, this way of putting things distorts what is really at issue by grossly understating the problem. It is not simply that we want to know whether we have any grounds for generalizing from the past to the future, or whether we have any grounds for thinking the future will be like the past. Even if we had an assurance that the future will resemble the past, this would not tell us *how* the future would resemble the past. Goodman's 'new riddle of induction', as his point is called, focuses on the difficulty of knowing how precisely to move from the past to the future.

The point Goodman makes is that the way you generalize from the past to the future will depend crucially on the categories you are using to classify things in the world. If, for example, you are looking at plants in terms of their shapes and you find that all the clover you have examined has three leaves, you will naturally infer that clover in the future will be three-leaved. If, on the other hand, you were not interested in its leaves, or if you did not have the concept 'three', you would not be likely to make such an inference, even given that you had a lot of experience of exclusively three-leaved clover (though without realizing it). Goodman, though, does not just point to the way we will generalize on the basis of the categories we actually have. He also argues that two sets of people could approach exactly the same evidence with different categories and come to different conclusions about future examples of the same type of thing. Their having different categories will, in other words, lead them to have different expectations about the future, on the basis of identical evidence.

Goodman's own admittedly rather contrived example concerns the colour of emeralds. It may seem that our past experience of emeralds would justify our claiming that all emeralds are green (or would if we accept any inductive inference). All the many emeralds we have seen have been green, and none not-green; so we have some good reason for thinking all emeralds are green. But, says Goodman, the same evidence exactly would justify us saying that all emeralds are grue. 'Grue' is a new predicate introduced by Goodman, and it applies to all things examined before the year 2000 just in case they are green, but to other things just in case they are blue. That is to say, a grue object examined after 2000 will be what we call blue, although grue objects examined before 2000 will be what we call green. Now, it follows from the definition and our evidence concerning emeralds that all emeralds we have examined are grue because something described as grue before the year 2000 will look the same as a green thing. Of course, on the same evidence, all emeralds so far examined are also green, but it cannot be the case that *all* emeralds are green *and* grue, because grue things examined after 2000 will look different from green things.

Goodman's example has been the subject of a vast literature, and numerous attempts have been made to show that there is something spurious about 'grue', often on the grounds that it is an unnatural sort of predicate, containing as it does a time reference. The standard reply to this is that, for 'grue' speakers, there will be something 'unnatural' about our 'green' and 'blue', which will have to be explained by means of a time reference. (For them 'green' = if examined before 2000 and found to be grue, or otherwise what they might call 'bleen', an amalgam of our blue and green ordered in the opposite way to that in which they are ordered in grue.) While this reply is adequate to the objection, over-

concentration on Goodman's own examples misses the substance of his point, which is actually another consequence of
one we made in discussing Bacon's presuppositionless observation: that you can generalize from a given body of evidence
in any number of ways. Looking at and generalizing about
our green emeralds under the concept 'grue' will lead you to
have quite different expectations about emeralds examined
after the year 2000, from those you would have if you looked
at them and generalized about them as examples of green
things.

At the end of his original lecture on 'the new riddle of
induction', Goodman said 'To say that valid predictions are
those based on past regularities, without being able to say
which regularities, is thus quite pointless. Regularities are
where you find them, and you can find them anywhere ...
Hume's failure to recognize and deal with this problem has
been shared even by his most recent successors.'[10] A grue
speaker in a grue world would, after the year 2000, see
regularities where we only saw differences. In our own
world, octopuses can apparently recognize shapes we do not
naturally perceive at all. This means that they will group
things and generalize about the world in ways that seem
bizarre to us. So one will project from the same evidence in
different ways, depending on the way one looks at the
evidence. Goodman's point is that general assurance that the
future will be like the past is not enough unless we know *how*
the future will be like the past, and *how* we are to project
from our evidence.

And at this point we can return from apparently abstract
and remote philosophical argumentation to the situation as it
is in science. Even though we have found a vast amount of
evidence confirming a given hypothesis, it does not follow
that the hypothesis is true. The next relevant observation

---

[10] In Goodman's *Problems and Projects* (Bobbs-Merrill, Indianapolis and New
York, 1972), p. 388.

may show it to be false. Europeans observed any number of
white swans, and the theory that all swans were white was
confirmed over and over again. But the first sighting of a
black swan in Australia showed the theory to be false. You
need only one piece of counter-evidence to disprove a general
theory, despite all its confirmations. This is the basic lesson
Hume would teach us. Then again, Newton's theories were
confirmed over and over again, but investigations of
Einsteinian relativity have shown that they are actually false,
at best only approximations to the truth in certain conditions. Here Goodman's point becomes relevant, because
the regularities in nature correctly pointed to by Newton's
theories have not been denied by Einstein. They have largely
been incorporated into the Einsteinian perspective, which
generalizes and projects from them in a different way, seeing
them as instances of different general laws from those envisaged by Newton.

Despite the difficulties for inductive proof raised initially
by Hume, and then by Goodman, there have been and
continue to be many attempts to work out systems of
inductive proof, showing the amount of support particular
bits of evidence lend to given hypotheses. But considered as
proofs, without further assumptions in the background,
when they are applied to the theories of science all these
systems seem to fall foul of the objections we have been
considering. They all seem to assume that the evidence we
have is likely to be genuinely representative of the whole
domain to which the theory applies and in the way the theory
generalizes from it. But it is just these assumptions that we
cannot take for granted in science. Our theory about swans
held only for the northern hemisphere. Newton's theories
hold good only for systems moving relative to each other
with moderate velocities. The regularities thus revealed
should have been conceptualized differently, leading to different projections in high velocity systems. Our experience is
restricted to a comparatively limited segment of space and

time, and we cannot safely assume either it or the way we look at it are representative of the whole. Nor can we simply assume that what we have experienced in our area of space and time will continue even there. Conditions may alter close to home, so as to refute our well-founded expectations even about our own locality.

Of course, we do project on the basis of our concepts and our past experience, and we have to do this. However, we ought to realize the assumptions we are making in doing this. It may be true that we are making a rather different type of assumption, and presumably taking a smaller risk, when we project a theory on the basis of past experience to similar conditions in the future than when we project the same theory into different types of conditions. But the difficulty is to know in advance *when* we are projecting into significantly different conditions, that is, which will make a difference to a hitherto exceptionless regularity. As far as swans are concerned, why should the southern hemisphere make a difference? And why, from a Newtonian perspective, should relative velocity affect physical magnitudes? And who could have foreseen that differences in just these respects would have amounted to a significant change in conditions?

The problems with inductive proof are compounded by what has come to be known as the paradox of confirmation, which can easily be illustrated by a simple example. Take the statement 'All ravens are black'. We learn from logic that this statement is equivalent to the two further statements 'All non-black things are non-ravens' and 'Any particular thing is not a raven and/or is black'. On the assumption that any evidence which supports or confirms a statement logically equivalent to another statement equally confirms that other statement, 'All ravens are black' will be confirmed by, for example, a blue car or a black violin, which confirm the second and third statement respectively. The supposed paradox is that a statement about all ravens being black could be confirmed by blue cars or black violins.

The paradox of confirmation is clearly not a formal paradox. Is it even really so surprising? Some would argue that it is not, given that evidence e confirms hypothesis h simply means that the information in e 'bears out' h. Certainly there is nothing about blue cars or black violins which throws any doubt on the blackness of ravens. In this sense, some bits of the world which are not ravens may be said to 'bear out' the hypothesis that all ravens are black. But this seems so cheap a way of getting positive confirming evidence for our hypothesis as to be practically speaking worthless, if we are really interested in putting our hypothesis to the test.

Faced with this difficulty and with the other difficulties confronting attempts to prove theories true, one reaction has been to shift attention from the notions of proof and confirmation altogether. After all, if any irrelevant bit of information 'confirms' a theory, and if even the best-confirmed theory is liable to be overthrown by the very next observation we make of one of its predictions, what is the value to a theory of masses of confirming evidence? Perhaps what really counts in science is not a mass of confirming or so-called confirming evidence—which might actually count for very little—but the serious attempt to probe a theory at its weak points. This proposal would actually have been rather attractive to Bacon, who thought negative evidence rather more important in assessing a theory than confirming evidence. When it came to matters of proof, Bacon was not a straightforward inductivist. In this chapter we have been highly critical of Baconian methodology, yet our analysis of inductive proof has shown that in one respect, at least, his intuitions were correct. Whether a stress on the negative testing of theories can avoid all taint of inductivism and allow us to set aside all the problems faced by attempts at inductive proof is what will concern us in the next chapter.

3
Falsification

Faced with the difficulties posed by the inductive method and inductive proof, it might seem better to make a fresh start and to think of science in rather different terms. It is the genius of Karl Popper that he has been able to provide a fresh start which promises to deal at the same time with the problems inherent in inductive method and in inductive proof. Popper's vision of the scientist is of one who uses his imagination freely and creatively, in order to produce bold and far-ranging theories, such as those of Kepler and Newton, which are then tested as severely as possible against the way the world is, and discarded if found wanting. The true scientist does not attempt fruitlessly to prove or make theories probable by the laborious piling up of insignificant and ultimately unavailing 'confirming' evidence. Rather, in the spirit of natural selection, scientific theories have to prove their mettle against the fiercest competition that can be found, and are allowed to survive only as long as they are not found wanting. And the community of scientists is seen as an ideally open society, in which anyone may propose ideas and theories, and anyone may criticize. In the Popperian vision, all are seekers after truth and all recognize the extent of their ignorance and the uncertainty of their knowledge. In the light of the ignorance and uncertainty which attends all human enquiry, the attempt to prove is displaced by the attempt to disprove, and the inductive hope that presuppositions can be eliminated from scientific work abandoned in favour of a full recognition of the role of the creative intuition in scientific research.

In the various accounts he has given of his intellectual development, Popper says that as a young man he was greatly impressed by the different attitudes of Einstein and of Marxists and psychoanalysts to their theories. Einstein made a bold and highly improbable prediction in his theory of gravitation, about the path of light being bent by the presence of a heavy body just as material bodies could be. This prediction was not tested until 1919, some years after having been made, when a total eclipse of the sun made testing possible. According to Popper, had this test gone against the theory, Einstein would have renounced his theory. By contrast, Popper says, Marxists and psychoanalysts always stressed the evidence in favour of their theories, and ignored or explained away counter-evidence. What impressed him was the contrast between the search for refutations manifested by the true scientists and the search for confirmations combined with a disregard for counter-evidence on the part of the others, and this led him to demarcate science from non-science in terms of truly scientific theories being exposed by their proponents to the risk of falsification or disproof.

We shall return to Popper's demarcation criterion in the next chapter. In stressing the falsifiability of scientific theories, Popper is drawing attention to the asymmetry between proof and disproof of a universal theory. As we saw in the last chapter, a universal theory saying that all planets have elliptical orbits can never be conclusively proved. We can never observe all planets, or see whether they have always had elliptical orbits or will continue to do so. We can never even know whether it is probable that the next planet to be discovered will have an elliptical orbit, even if all observations up to now are of planets with elliptical orbits. On the other hand, it is clear that only one case of a planet not having an elliptical orbit is enough to refute the theory that *all* planets have elliptical orbits.

Here it seems is something science can get its teeth into, whilst allowing for theories which go far beyond existing evidence. We cannot prove theories, but we can disprove them. Hence what we ought to do in science is not unavailingly to look for proofs of our theories by accumulating empirical evidence in their favour, but rather we should attempt to disprove our theories, and to aid the process of disproof by constructing theories which make bold and unexpected predictions, for such theories would, if false, be easy to disprove. And it seemed to Popper that this method of conjecture and refutation correctly described the history of modern science. In particular, it accounted for the overthrow of Newtonian physics despite its previous unprecedented successes. Popper says that with his theory of science as a process of falsifying and attempting to falsify theories he has solved the age-old problem of induction. But what he has done is not so much to solve the problem of induction, as to side-step it altogether by regarding science in terms of disproof and by suggesting that, in science, proof of an inductive sort has no role to play at all.

If Popper had simply said that scientists should imaginatively propose bold theories, which they then try as vigorously as possible to overthrow, it is unlikely that he would ever have been taken very seriously. While the proposing and testing of theories is undoubtedly an important part of the scientific enterprise, it is only a part. We also naturally expect science to increase our positive knowledge and to give us theories we can rely on for practical purposes. Popper does indeed have things to say on these points, and these go some way to making his account more generally palatable. Unfortunately for him, it also tends to make his claim to be able to do science without inductive assumptions considerably less plausible.

Although according to Popper we cannot positively prove

or confirm a scientific theory, we can sometimes speak of a theory as being well-corroborated. A theory is well-corroborated if it is highly testable and if it survives severe testing. The testability of a theory is related to its capacity for yielding testable predictions and to its degree of empirical content. Roughly, the idea of high empirical content is that a theory which is simple, bold, and highly precise is very likely to be false, and is thus more testable than a theory more hedged about with qualifications and exceptions. The bold theory has more content in that in an obvious sense it forbids more. It says more about how the world actually is, and thus is more likely to be found false by the world. Severe testing of a theory comes when a theory makes predictions which are highly improbable in the light of our existing knowledge—as Einstein's did in predicting the bending of light. Indeed the solar-eclipse expedition and observations may be taken as a paradigm case of an attempt at the severe testing of a theory, which also, incidentally, had high empirical content. That theory survived its severe test, and could thus be accorded a high degree of corroboration, relative to other high-content competitor theories which either had not been so severely tested or had failed their tests.

But why, one wonders, should anyone be interested in degrees of corroboration? Popper is insistent that a degree of corroboration is always relative to the success of other theories and always backward-looking, being simply a report on past performance. He has to say this, of course, or his theory will be covertly inductive. If he were to say that the fact that a theory *had* survived severe testing was a positive reason for our believing it to be true or likely to be successful in the future, then he would clearly be using degree of corroboration as if it were a measure of positive (inductive) confirmation.

Let us suppose for a moment that we rest content with the corroboration of a theory as being, as Popper says, simply a measure of its past performance, and not being covertly thought of as a pointer to its future success. Will we have succeeded in eliminating all taint of inductive argument from our methodology? That we should do this is very important for Popper. He fully accepts Hume's scepticism about induction and believes that he has succeeded in formulating an account of science which is free from all taint of inductivism, and which, at the same time, shows science as an activity guided by reason rather than by irrational forces, and one which we may hope is all the time increasing our knowledge. The first problem for a Popperian to consider, though, is whether he can really talk of a severe test without the use of inductive reasoning.

Popper is insistent, rightly no doubt, that repeating an old test of a theory does little to enhance its degree of corroboration. Nor does it increase its degree of corroboration if it survives tests whose outcome we would already have expected to support the theory before the theory had been formulated. There is both a law of diminishing returns in testing a theory and an expectation that genuine tests will not simply confront the theory with evidence we already had, but will provide it with an independent test. We expect a new theory to prove its mettle by anticipating knowledge we did not already have, as relativity theory did with the bending of light. A severe test of a theory is only possible when a theory makes an unexpected prediction, one that is unlikely or surprising on present evidence.

All this is sound good sense, and a useful corrective to the idea we find in some inductive logicians that every new piece of confirming evidence, however well-known or expected it might be, adds significantly to the probability of the theory. But it is unclear that it is good sense Popper is entitled to. For a severe test is one which is *unlikely* on past evidence. Without using some sort of inductive assumptions, how can one move from past experience to calculations of present (or future) probability? Looking at yet another swan in the lakes of Regent's Park in London may not, one would probably feel, be a severe test of the theory that all swans are white. Yet Regent's Park is just where you could put the theory to the test. (There are some black swans there.) And what, in any case, could a non-inductivist have to say about the probability or improbability of the next swan to be observed in London being white? All we have, on non-inductive grounds, are reports of *past* experience, and generalization from them is forbidden. Similarly, up to 1919, the bending of light may not have been observed. But what reason had a non-inductivist to believe that in 1919 light might not start bending? The crucial point is that it is only against a background of expectations built up from past experience that we can speak of some outcomes being improbable, and hence of severe tests in the Popperian sense. Without some sort of inductive argument, all tests are liable to look equally severe.

But even if we could know non-inductively that a given test is severe, why should we be at all interested in conducting severe tests on theories? One of the things we want to establish in science is some idea of the basis on which we might act in the future. Popper is reluctant to see the manipulation or control of nature as a main aim of science; he is Aristotelian rather than Baconian in that respect. Nevertheless, he does admit that what he calls a 'pragmatic' preference for one theory over another is something we can legitimately derive from science; given that we have to act, we should act on the most rational assumptions we can. There is no method more rational than that of the proposing of bold theories, and their severe testing. We should, therefore, act on the results of this method.

The obvious rejoinder to this is that a rational method for sifting theories with regard to their past performance is not at all the same as a rational method for sifting theories with regard to their future performance. For that we would need something like an inductive jump, from past to future. And it is just that which Popper believes we can do without in science.

At this point the dispute between Popperians and their critics often begins to take on a slightly unreal air. The Popperian will say that we have no reason for not acting on the best tested theory, while the critic will say that we need something more than 'no reason against'. The Popperian will say that we have a reason against acting on certain theories, such as the theory that descending high buildings via the window is safer than using the stairs, in that they have been refuted, while the theory that the stairs are better is as yet unrefuted. The critic will say that this knowledge concerning past refutations provides reasons for future actions only on inductive assumptions; without such assumptions there is no reason against jumping out of the window; and so we are back where we started.

The Popperian will say that the Popperian method aims at the truth. The critic will reply that the method aims at the truth only in the sense of ruling out false theories, and that it does not give any positive reasons for believing in the theories which have survived severe tests. To which the Popperian will agree, adding that we may still act on such theories in the hope that they are true. And the critic will say that he had hoped for more than a hope in science. Once again, we are back where we started.

A Bayesian Approach

What makes discussions between Popperians and their critics more than a little surrealistic is that they can seem largely verbal. The critic and the Popperian will often agree on which theories are the best tested and on which one should act, and on which consequences of a new theory are improbable. On substantive matters, it seems, there is a measure of agreement. Disagreement arises mainly on the description and analysis of what is going on under the surface, yet even this disagreement can be more apparent than real. Unless the critic is hopelessly benighted about induction, he will agree that past experience and severe testing do not guarantee future success, and he will agree that the mere piling up of similar confirmations of a theory does little to increase its probability. And in fact, as we will see he does have a somewhat better account of all this than Popper can give.

For Popper, the growth of knowledge in science is a basic premiss, and his philosophy is largely an attempt to analyse this. But what can growth of knowledge consist in for the Popperian? From the Popperian point of view, the history of science will look like a progression from falsified theories to false theories which are as yet unfalsified. The only success we can have good reason to think we get in science is the falsifying of a theory. On the way, we may pick up bits of observational knowledge, such as sightings of new planets, physical measurements, knowledge of the bending of light, and so on. The real Popperian aim for science is the gaining of true universal theories about nature. As far as that is concerned, the growth of knowledge is largely a growth of knowledge of which theories are dead. The history of science for the Popperian is the graveyard of decreased theories; to vary the metaphor, the aim of science looks like the systematic suppression of inept species. Indeed, this points to perhaps the most curious feature of all in Popper's philosophy of science, where there is a real difference between Popperians and others: the insistence that in science we are actually to prefer theories which are less probable and more likely to be false to those which are (relative to background knowledge) probable and likely to be true. This insistence makes sense only on the assumption that the real aim of science is to eliminate false theories. For Popper, the best theory at any time is not the best-tested theory, but rather the one which has most potential for future falsification. As he puts it in an addendum to the 1972 edition of *The Logic of Scientific Discovery*: 'By the "best" theory I mean the one of the competing and surviving theories which has the greatest explanatory power, content, and simplicity, and is least *ad hoc*. It will also be the best testable theory, but the best theory in this sense need not always be the best tested theory.'[1]

While we can admit that the elimination of false theories and the invention of wild and improbable theories both have a role to play in science, it would clearly be an exaggeration to think of the aim of science exclusively in these terms. We also expect science to give us theories that have some positive chance of being true, by virtue of a high degree of evidential support, and which we can rely on. Popper's insistence on the best theory at any time being the most *testable* theory shows the extent to which he conceives of science as a theoretical enterprise, detached from its grounding in and continuity with our everyday factual beliefs and unrelated to such mundane matters as its practical applications or its role in benefiting man's estate. It is surely significant that when speaking of the aim of science, Popper always tends to speak in terms of *explanations* of phenomena in *universal* theories. But once again, we have to insist that proposing and testing universal theories is only part of the aim of science. There may be no true universal theories, owing to conditions differing markedly through time and space; this is a possibility we cannot overlook. But even if this were so, science could still fulfil many of its aims in giving us knowledge and true predictions about conditions in and around our spatio-temporal niche.

The inductivist who is aware of the dangers and risk of failure in generalizing from past experience and conceptualizations to events and objects of which we have no experience

---

[1] *The Logic of Scientific Discovery* (London, Hutchinson, 1972 edn.), p. 419.

will take a rather different tack. He will begin by emphasizing the extent to which life of any sort and the acquisition of knowledge depends on the existence of a relatively stable environment, and of a degree of adaptation between the environment and the organisms that live in it. There is nothing guaranteed here, as Hume and Popper insist. Our expectations could let us down. The environment could suddenly change, or we could find ourselves in an environment to which we are unsuited. Moreover, as Goodman's new riddle of induction shows, the regularities our concepts lead us to project into the future may not be, objectively speaking, projectible at all. Our concepts themselves may, in that sense, be burdened with false theoretical assumptions, derived no doubt from their past successes in helping us to cope with the environment.

Against these sceptical difficulties, however, we can ask ourselves how, given that there really are some regularities in our environment, we can set about discovering them. We will naturally have to start from the dispositions and primitive theories we are given in our sensory apparatus and the language we learn as children. Not only do we have nowhere else to start from, but the regularities indicated there are not pure figments of imagination. They have stood the test of time, and generalizing on their basis has at least not prevented the survival of generations. Given that our environment does not radically change, we have good reason to accept them as a starting-point.

But we also know that the regularities postulated in our sense apparatus and first language may not be projectible. They may not have much direct connection with the underlying structure of our world; they may be based only on the way things appear to us or on things which behave in similar ways only in certain special conditions and not because they form a genuine natural kind; and they may also only apply in a highly limited area of space and time. We will naturally want to seek regularities at a more fundamental level of nature than at that of common-sense observation, and which, for that very reason, have a good chance of applying beyond our region of space and time.

To take an example of what I mean here, colour is a pervasive feature of the world as observed by us. But we now know that, scientifically speaking, colour is not a fundamental property of the world. It is a by-product of more fundamental processes acting on our sense-organs. Just how we interpret this fact is an open question, but it nevertheless serves to illustrate how a perceptually dominant feature of the world—and the regularities based upon it—may be explicable in terms of scientifically and causally more fundamental properties, such as the activities of (colourless) photons on our retinal surfaces, and the like. Similarly, to use Roger Scruton's example, stonemasons refer to porphyry, onyx, and marble itself as ornamental marbles. In appearance, they are all similar and this similarity is exploited in building. But they do not form a natural kind. Porphyry is a silicate, onyx an oxide, and marble itself a carbonate. In plenty of circumstances their different underlying structure would lead them to behave quite differently.

How, though, might we hope to arrive at theories of a greater depth and breadth than are given in our initial predispositions and to find genuine natural kinds, united by similarity of deep structure, which would lead them to behave similarly in all sorts of as yet unencountered situations? The way forward here is to propose theories which predict effects which go beyond what we are already aware of, in the hope that we might uncover both particular facts and general regularities in such areas. This is what actually happens in modern science, where the theories proposed lead to predictions, and hence to observations about phenomena both *beneath* and *beyond* the effects we observe in our visible world. The theories of physics, chemistry, and biology all take us to the underlying structures of their respective domains and to phenomena beyond those of which we have experience. Theories which survive severe testing in areas outside our experience may be good indicators of regularities more fundamental and wide-ranging than those of which we do have experience and can form the basis for better future projections than our initial ones. At least, if there are any such regularities and any genuine natural kinds underlying such regularities, there seems no better way of trying to reveal them than by proposing theories purporting to describe them, and then testing these in new areas of experience, to see how things might be in areas about which we originally knew nothing.

As far as method goes, much of what I have just been outlining would be like the Popperian method. The stress on the positive role for severe testing, and going beyond what we already know, is common to both accounts, as is the tendency to downgrade the force of repeated observations of regularities which are already known. For such regularities, it is conceded on both accounts, may be of very limited application and significance. They may be quite local, due to some temporary coincidence of causes, which just happen to have come together in special circumstances without being based in a genuine natural kind or being generally projectible or supportive of counter-factual statements about what would have happened had circumstances been different. On the other hand, the model I have just proposed does not downgrade existing knowledge in the way Popper does or think it of no probative significance at all. It is regarded as the point from which we start, and from which we can go further. And, with the assumption that we do live in a world of some stability of which we have some knowledge, this is surely correct. The world might suddenly and radically change, and if it did most of our present knowledge and conceptualizations would be quite useless. But that is no reason for not making the attempt to gain as much knowledge as possible about the world as it now is, or for not aiming at a judicious interplay between our old knowledge and the proposing and testing of new theories beyond our old knowledge, which may correct it, or show it to be part of a wider and more inclusive whole. Such interplay can give us good reasons for thinking we can find out quite a lot about the world as it now is, and will continue to be, on the assumption of comparative stability. Indeed, one thing our deeper and wider theories can importantly reveal is when a little local regularity is likely to fail us—against the background, of course, of a deeper and wider one. And this is something we might be led to see through working on theories which aim to correct our old knowledge and our old projections; we might explain it as the result of more fundamental and wide-ranging features of the world.

Much of what I have tried informally to capture about science and its theories in this section is expressed formally through Bayes's theorem, a formula which, given certain assumptions, follows from the probability calculus and which describes the probability of a theory after some test evidence has favoured it. The upshot of Bayes's theorem is that this probability is increased *both* by the severity of the favourable test evidence *and* the initial probability of the theory being tested.

Let us take h to be the theory we are testing, e the evidence of our test, and k our background knowledge prior to the test. What we want to know is the extent to which the test might be said to confirm or make the theory probable (given that the test goes in favour of the theory). We write this measure as

$$P\ (h/e.k)$$

that is, the probability of the theory, given both our new test evidence and our background knowledge. Then Bayes's theorem has it that this probability *increases*, the severer the test is, given our existing background knowledge: if, as with Eddington's observation of the total eclipse of the sun in 1919, the predicted outcome is unlikely on our existing knowledge and the test therefore severe on our existing knowledge, a favourable outcome increases the probability of the theory. But the theory's probability *decreases* if the theory itself is highly improbable on our existing background knowledge. Formally, in one version, Bayes's theorem states

$$P(h/e.k) = \frac{P(e/h.k) \times P(h/k)}{P(e/k)}$$

In testing a universal scientific theory $P(e/h.k)$ will generally be 1, as the test evidence e will normally be predicted by the theory h, so it is easy to see that $P(h/e.k)$ will increase, the lower $P(e/k)$ is and the higher $P(h/k)$ is.

Bayes's theorem, then, would suggest that in science we should seek theories which have some prior probability relative to what we already know and then attempt to test them severely. In line with Popperian corroboration, a theory could not on Bayesian grounds receive any significant increment of probability if it could not be severely tested. Nor would it gain much probability from a test whose results would be predicted on background evidence alone, apart from the theory. So we are encouraged to do more in our theories than simply generalize on the basis of existing data. On the other hand, against Popper, implausible theories are not in themselves good. Our theorizing is to be constrained in a positive way by what we already know, if we want theories that are regarded as highly probable on Bayesian grounds.

It may be objected that Bayes's theorem expects us to assign a degree of initial or prior probability to theories and evidence relative to background knowledge, and that this is a highly subjective matter. There will thus be an ineradicable subjectivity in calculating the posterior probability of a theory after the evidence is in, owing to varying personal estimates of the prior probabilities. Against this, though, some would argue, following the work of de Finetti and Savage, that initially different prior probability assessments

tend to converge, with increasing amounts of relevant evidence. And this convergence would seem to reflect what
would be desirable practice. We expect assignment of the
degree of credibility of theories to firm up on increasing
evidence.

It must be recognized that quantitatively to work out
degrees of probability or confirmation or even corroboration
for actual scientific theories is well-nigh impossible. Those
who have studied such matters have tended to restrict their
efforts to simple formal languages because of the problems
involved in dealing with the logically more complex languages of science. Nor are there precise ways of estimating
and relating the various different features of a theory which
might have a bearing on its probability, prior and posterior,
or indeed of the variety of types of evidence which might be
adduced in its favour, or of the relative severity of different
types of test. Our discussion of Bayes's theorem should not
lead anyone to suppose that there is or could be real rigour
here. Despite this, Bayes's theorem remains a neat way of
suggesting that severe testability is not all we want in a
scientific theory. We do want that but, on the assumption
that we are conducting our scientific investigations in a
relatively stable world (or a relatively stable corner of the
world), we also want theories which have some reasonable
initial probability relative to what we already know.

In practice, Popper would probably admit much of this.
At least he thinks that an acceptable future theory must
account for the past evidence we have. Nevertheless, the
implication in his account of corroboration that we cannot
give a theory any positive credit for doing this is surely
wrong. It is in line with his anti-inductivist stance, but that
stance is itself thoroughly suspect. Not only is it hard to spell
out the concept of a severe test non-inductively, but Popper's
philosophy of science gives us no relevant reason for acting
on previously well-corroborated theories. Once we see our
scientific activity against the postulation of a relatively stable

environment of which we have some knowledge, and within which we may have reason to think we have uncovered some natural kinds, we can see at once why we do have some reason for acting on the best-tested theory, for such a theory is the most likely to have uncovered genuine and wide-ranging regularities. And we can also see why initial probability in a theory is to its credit, for such a theory will be based on those regularities which have characterized what we so far know of the world. None of this should be read as 'proving' what cannot be proved: that the future will resemble the past in just the ways we are led to expect by our theories. There can be no direct refutation of Hume or Goodman in that way. But that does not mean that there is no perspective from which severely testing our theories and relying in the future on past evidence, especially from severely tested theories, might not be an eminently reasonable course of action, and so far in this section I have tried to sketch such a perspective.

In speaking of the reasonableness of severe testing and of relying on theories which survive severe tests, I have not, of course, said that what we are thus led to postulate about the world is shown to be true. Equally, I have tended to speak of the discovery of natural kinds and of far-reaching regularities in the world as if they were two sides of the same coin. Many, though, would put the aim of science in rather stronger terms. For them science is primarily about the discovery of the essences of things and of physical or causal necessities which would determine not only the actual regularities in the world, but also things that would have happened in circumstances that did not actually come about. On this stronger view, to speak of a natural kind is not simply to speak of a group of things which have a common underlying structure and just happen to behave in similar ways. The stronger view of science would have it that, because of their underlying structure, members of a particular natural kind have to behave in the way they do, and would have behaved similarly even in counter-factual circumstances (that is, ones which did not actually occur). It is easy to see that if this line is taken, an anti-Humean line will also be taken on induction. If the nature of things means that they are bound to behave in certain ways, certain future events would of necessity follow certain past events, although, as we will see, it is another question as to whether we would ever be justified in asserting this of any two events.

The notions of physical essence and causal necessity are connected in that the physical essences of things would be held to underlie the relevant causal necessities. An acceptance of these notions will go hand in hand with an anti-Humean view of induction and with the view that science is about more than the discovery of regularity in the physical world. On the stronger view, it is about that, but at least some of the regularity in the world will be taken to be indicative of something deeper, of the very essence of the world, of something which makes regularities happen the way they do and which would guide what potentially happens in circumstances that do not actually come about. There is no denying the pervasiveness of this type of thinking in our lives, and indeed its persuasiveness. If we notice wide-ranging regularities in phenomena we naturally feel that there is likely to be some deep explanation, in terms of the essences of things, which means they have to behave as they do. Moreover, we all recognize that there is a fundamental difference between regularities due to physical law and those due to accidental concatenations of circumstances, between the regular falling of heavy bodies to the earth, say, and the equally regular, but presumably quite accidental phenomenon that all gold bars have a weight of less than 100,000 tons. Science deals with the essential regularities in nature, not the accidental, and according to the stronger view, it distinguishes between the two in terms of what is contingent and of what belongs to the nature of things and is necessarily connected. Indeed, part of the reason for making the distinction between a genuine causal regularity and a merely accidental one is the fact that in the former case, but not the latter, we have a theory which attributes the causal regularity to some underlying and necessitating feature of the world, such as a force governing the deep structure of things.

The world may indeed have a deep structure and things may have essences necessitating their behaviour. If this is so, then the methods of science—of theorizing on the basis of and at the same time beyond our experience and testing those theories—will offer us the best chance of uncovering what has to be in the world. But the lesson of the scepticism of Hume and Goodman unfortunately remains: even if there are natural necessities and strongly necessitated natural kinds, we cannot ever be sure that our theories and concepts are actually discovering them or describing them correctly. We cannot be sure that our science is fulfilling this fundamental aim. As we have seen in this chapter and will see further in Chapters 5 and 6, the most worrying problems in the philosophy of science are fundamentally epistemological. We can adumbrate easily enough an anti-Humean metaphysics, a vision of the world as ruled by causal necessity and inductive regularities and within which we can securely speak of what would have happened, had circumstances altered. But that does not license us to think that *our* theories have uncovered the truth about any of these things, or that what our theories uncover have any application beyond our limited experience.

In so far as we are entitled to make assertions on the basis of our theories and the evidence they are grounded on, all we know is that certain regularities have obtained in our experience of the natural world. We have no guarantee that they will continue to obtain, or that their obtaining is due to any real necessities our theories have revealed. The explanations given by our theories may be erroneous, and the regularities they explain be due to some other cause, or even to no further cause at all. The regularities may just occur. This is a possibility our evidence does not entitle us to rule out, and, as we will see in more detail in Chapter 6, some influential philosophers of science would analyse the achievements of science in terms of the tabulations of observable regularities in nature rather than in terms of the revealing of deep necessities and essences underlying those regularities. What remains on either interpretation of science is that there is merit in a Bayesian approach to the assessment of theories. Whether we consider our theories as simply revealing wide-ranging regularities in nature, or whether we conceive of them as disclosing the deep structure of the world, the theories that will be most acceptable will be those which build on but lead us significantly beyond current knowledge, for by means of such theories we may hope to unearth wider-ranging regularities in nature than those of which we already have knowledge—if there are any. And it is only on the basis of pervasive regularities in nature that we have any hope of uncovering natural necessities and the essences of things—if there are any.

4

Science and Non-Science

The Demarcation Criterion

In the first chapter, we suggested that there has been a type of growth of knowledge in science which has not been found elsewhere in human thought. It was also suggested that in science proper there was an attention to criticism and the negative instance which was lacking in certain other areas of intellectual activity, rather to the detriment of those areas. These two points taken together suggest that there might be some point in making precise our intuitive notions of what science is, in order to frame what has become known as a demarcation criterion between science and non-science. In this way, we might hope to understand both the progressive nature of science and its intellectual prestige.

In a way, of course, Bacon was attempting to formulate just such a demarcation criterion. True science was to consist in the meticulous analysis of masses of presuppositionless data. This would guarantee growth of knowledge and underwrite the honesty and intellectual integrity of the discipline. Unfortunately, though, Bacon's methods are unworkable, and his description of the formulation of scientific theories is a caricature. The failure of Bacon's ideas might suggest that the scientific spirit consists not in the way we formulate our theories, so much as in our treatment of them once we have got them. Our presuppositions are always with us, never more so than when we think we are doing without them. Let us accept this fact together with the role of creative insight in scientific thought. Science then will gain its distinctive character not from the elimination of presupposition and intuition, but in the control an impartial nature will exercise over them.

What is being urged here is a distinction which is simple enough to grasp on the surface but apparently very hard for many critics of scientific activity to accept in practice. This distinction is sometimes misleadingly referred to as that between the context of discovery and the context of justification. The idea is that in the formulation of a scientific theory—the so-called context of discovery—anything and everything may be thrown in: Kepler's mysticism, Newton's alchemy, Kekule's dreams, Haldane's politics, Keynes's moral views. There can be a million and one influences, intellectual, financial, emotional, social, cultural, political, subjective, and objective, which lead scientists to come up with the sort of thoughts they do. The context of discovery is quite uncontrolled—which is good because otherwise we would be stuck with the same old thought processes and never gain new perspectives. But what is being spoken of here is not really the context of *discovery*. It is rather the context of hypothesis-formation.

We only get to discovery, if at all, at the next stage, the so-called context of justification, when the theory that is proposed is shaped and formulated so that it can be tested, and actually tested against nature. And what we have here is not strictly a context of *justification* for the theory as a whole, because we cannot justify it. The evidence of nature can never show that a theory is really true, but at most that it survives so far. What we have is like the context of a trial in a court of law, where you can be found guilty, but where being declared not guilty does not mean that you are justified before God, but only that the evidence does not condemn you.

The distinction between the context of discovery and the context of justification appears to be difficult to grasp. Many are so impressed by the influence of social and historical context on scientific work that they fail to see how this context is significant only to the context of discovery. Even if, as is sometimes claimed, the spirit of capitalism created a climate in which men would naturally seek to quantify, analyse, and exploit nature, it does not follow that all the theories produced in this context are not true. Whether they are found wanting or not will be determined by their ability to predict the course of nature, and not by the desires or beliefs of capitalists, entrepreneurs, mystics, or social historians. And the same goes for research initially inspired and funded by military sources. Scientists may be researching into lasers because of the Strategic Defence Initiative funding, but that has no bearing on the truth of any resulting discoveries or on the effectiveness (or ineffectiveness) of the outcome.

All this goes to suggest that it might be possible to take empirical falsifiability as the distinguishing mark of a scientific theory. Proposing falsifiable theories and actually testing them will control the context of discovery and the scientifically impure motivations and flights of fancy found therein. Weeding out falsified theories and suggesting improved ones will also give a reasonable hope that knowledge might grow. While this will not amount to a guarantee of actual growth of knowledge, the pressure will be on for researchers to expose their theories to the objective demands of nature and so extend our knowledge. And honesty and openness will be preserved by the impartial testing by nature of the creations of at times dishonest and highly partial men.

That science should be demarcated from the non-scientific in terms of empirical falsifiability is the proposal of Popper which has most captured the imagination of the general public, who have seen in it a means of justifying their suspicions of influential pseudo-sciences. And certainly a theory purporting to be about how the world is, but which is not testable by means of observation and experiment, will rightly be an object of suspicion. A theory compatible with many worlds will, to that extent, tell us little about the precise characteristics of this one. This indeed is the burden of Popper's hostility to psychoanalysis: that its theories are compatible with any type of behaviour in individuals. Whether you are brave or cowardly, you will still be manifesting your Adlerian will-to-power. A dutiful son and an adolescent rebel may both be suffering from an Oedipus complex. Phenomena of this sort have suggested to many— perhaps knowingly or unknowingly influenced by the Popperian demarcation criterion—that psychoanalysis does not provide explanations of a scientific sort, in which predictions are derived from theories, and then tested against the evidence, so much as persuasive redescriptions and interpretations of human activity from specific standpoints, looking at human activity as an instance of will-to-power or of repressed sexuality, for example. The success of such stories will presumably be judged by their adequacy in describing and making sense of the behaviour of the individuals in question, rather than in terms of any predictions they might lead to. The skill of the analyst will be more like that of a novelist than that of a scientist.

The case with Marxism is rather different. Here various predictions have been made. Revolutions will occur in industrialized, capitalist societies. In such societies there will be an increasing polarization between capitalists and proletarians, with the proletariat becoming more and more impoverished. Capitalism itself will reach a stage of terminal crisis, and conditions will be ripe for revolution. After the revolution, after a period of proletarian dictatorship, the state itself will wither away. None of these things have happened, of course. Marx did not foresee the rise of the administrative classes, nor the extension of the welfare state, nor did he foresee socialist revolutions in agrarian societies, nor understand the extreme reluctance of workers' states to wither away. Popper's reaction to all this is to ask why Marxists do not simply accept the falsity of Marxism. Instead they cling to it as to a religion, devising ever more arcane and complex explanations for the failure of its predictions. This in itself would, from a Popperian point of view, be reason enough for thinking Marxism is no longer a scientific theory, properly speaking. It may once have been. It did make predictions, but its proponents no longer treat it scientifically, taking its falsifications on the chin. Instead they weave and bob around the ring, ducking blows and slipping out of clinches. Yet, while not a science, like psychoanalysis it gains much of its prestige from being thought of as, in some way, scientific.

This takes us to the other side of the demarcation criterion. It is not just that we want to know in what science consists, in order for us to understand it. We also want to be able to distinguish the genuine article from the fake. For a Popperian, there is no necessary discredit in not being scientific. If science is defined in terms of the falsifiability of theory by empirical states of affairs, then mathematics—a paradigm of rationality—is not scientific. Neither are ethics, or literature, or music, or philosophy, and all of these are important and valued intellectual activities. What is discreditable is not being a non-science. It is pretending to practice science when you are really enacting a moral vision or a therapy or a religion. In these circumstances, a clear demarcation between science and pseudo-science would undoubtedly be helpful.

The difficulty with Popper's demarcation criterion is that matters are nowhere near as clear as they originally seem. At one point Popper says that 'statements or systems of statements, in order to be ranked as scientific, must be capable of conflicting with possible, or conceivable observations'.[1] This clear thesis, though, raises at least two immediate problems and a more fundamental one. First, many empirically provable statements, which we would intuitively think of as scientific, become unscientific. Thus, 'There is at least one

[1] *Conjectures and Refutations* (London, Routledge & Kegan Paul, 1969), p. 39.

planet', 'There are electrons', and 'Bacteria exist', are all unscientific, because unfalsifiable. We cannot disprove them by observation, in other words. This lack of conflict with actual or possible observation arises from their generality, from their saying in essence that somewhere or other in the universe things of such and such a type exist, and we can, of course, never examine the whole universe so as definitely to rule out the existence of the type of thing in question somewhere or at some time. The statements in question are what logicians call existential generalizations, and it is not implausible to think much of the knowledge we acquire in science is of this type: knowledge, that is, that things of such and such a type do exist somewhere in the universe. Obviously, statements of this sort can be proved, if somewhere we discover an atom or an electron or a bacterium, as has, of course, happened.

The second immediate problem concerns probability, which plays an increasingly important role in science. A probability statement is one which says that a particular proportion of events will be of such and such a character, but without specifying which ones. Thus, we can say that a certain coin has a 1 in 2 probability of coming down heads ($p(h) = 0.5$). The problem with such a statement is that it cannot be falsified if no limit is put on the possible number of coin tosses. 10,000 tails in succession would not strictly refute $p(h) = 0.5$, because over a very long run of tosses 10,000 tails might be balanced out by a large population of heads, and this could be said of any deviation at all from any predicted probability. The probabilistic theories used in science usually refer to open-ended runs of events, and so it looks as if they cannot be falsified.

As we shall see further in Chapter 7, Popper does, however, have an answer to this point. Statistically, certain runs (like our 10,000 heads) would be highly unlikely if $p(h) = 0.5$ were true, and this can be mathematically computed. So we can know in advance what the likelihood of certain outcomes would be, given the truth of a particular estimate, and we can then stipulate that certain outcomes highly unlikely on our theory are to be treated as refutations of that theory. There is, to be sure, a certain degree of stipulation and arbitrariness here, in that their unlikeliness does not logically entail the falsity of our hypothesis. None the less, as we shall see, there is a degree of stipulation about falsification as such, and certainly a move of this sort on probability statements would preserve the spirit of submitting scientific theories to the risk of empirical falsification.

Popper might get round the existential generalization problem by saying that truly scientific statements would have to be either disprovable or provable by observation and that probability statements can be treated as falsifiable, as we have just seen. Even so, his criterion would still face the more fundamental objection that many statements of the sort which he and everyone else would recognize as among the most important scientific statements are not actually capable of proof, as we have seen, or of conflicting with possible or conceivable observations, without certain additional assumptions.

Are Theories ever Falsified?

Let us take an example of a scientific statement, Newton's Third Law: for every applied force, there is an equal and opposite reaction. As it stands, this statement, like all statements of scientific theory, is entirely general and tells us about no specific state of affairs. In order for a theory to conflict with any specific observation, actual or possible, we will need to be given an actual instance of an applied force together with its magnitude. Then we will be able to predict that there will be an opposite reaction of the same magnitude. Thus, if two spring balances are hooked together and pulled, we will be able to predict that the one will register the same force as the other. If the force registered by the second was not equal to the force registered by the first, we would then appear to have a disconfirmation of Newton's Third Law. The first point to notice, then, is that a general theory needs to be combined with some relevant particular observation statement before any prediction is yielded. Further, the actual effect will also have to be observed and recorded in another observation statement before there is any falsification of the theory.

The significance of this role of observation statements in the actual testing and falsifying of a universal theory is that it is always possible to deflect criticism of a theory by questioning the truth of one or other of the observation statements involved in the testing. We may want to test the truth of the general theory 'All swans are white'. From 'All swans are white' and 'Cyrus is a swan' we may deduce 'Cyrus is white'. We may observe that Cyrus is black, and think we have a falsification of 'All swans are white'. But are we sure that Cyrus is a swan and that Cyrus is actually black? Cyrus may actually be a creature that only resembles a swan, or his blackness due to some disease or temporary discoloration. Avoidance of falsification in this sort of way may be rather implausible in the case of Cyrus the swan, but it cannot be ruled out on logical grounds. And in the case of the complicated and difficult observations and measurement involved in the testing of scientific theories, questioning the observation statements involved need not even be particularly far-fetched. Newton is reputed to have told the Astronomer-Royal Flamsteed to repeat his observations and recheck his results when they were found to have conflicted with the predictions of Newtonian theory, and Flamsteed, on doing so, to have admitted that he was originally mistaken.

Mention of the repetition of observational results brings us to another feature of scientific practice which makes the falsifying of theories more difficult and uncertain. A key feature of the objectivity of science is the repeatability of observations and experiments. Insisting on repeatability guards against observer bias and inaccuracy, to say nothing of dishonesty, and against freak results due to chance or unusual factors interfering with a particular observation. As such it is a crucial aspect of the objectivity and openness of science, but it means that theories are falsified not by single observations or experiments, but by what is in effect another general hypothesis to the effect that such and such an observation is repeatable. Once again, there is scope for the defender of a theory to claim that a particular result or set of results apparently refuting the theory are not examples of a genuinely repeatable effect.

A universal theory describing the behaviour of all ex-amples of a particular type of phenomenon will be logically proved false if a prediction which follows from the theory, in conjunction with a relevant observation statement, goes against the theory. But, in practice, falsification will not be accepted if there is doubt about any of the observations or their repeatability. And there is yet another respect in which scientific practice is more complicated than the logical situa-tion, which turns out to be the most unsatisfying of all for those who like things to be cut and dried in science. There is always an implicit assumption in any test situation that there is no factor interfering with the observed result so as materi-ally to affect the result. Thus, it would not actually count against Newton's Third Law if one of the spring balances in our test is failing to register the appropriate equal and opposite reaction because there is some defect in the mechanism of its needle. Nor was it any refutation of Newtonian theory when the planet Uranus was seen to have an orbit other than the one predicted from the application of Newtonian theory. It was no refutation because—as was predicted eventually—another hitherto unknown planet was exerting a gravitational pull on Uranus, and affecting its orbit. This prediction of a new planet was, of course, confirmed by the discovery of Neptune, and Newtonian theory in this instance triumphantly vindicated.

The discovery of Neptune was a case where postulating an interfering factor affecting the test outcome led at the same time to a positive growth of knowledge in the discovery of the new planet and to a vindication of the threatened theory. But things do not always work out so neatly, and this is where the uncertainties begin. It is always possible to postulate interfering factors to explain away the apparent failure of a theory in a test. If one explanation fails, another one can always be tried. In the last resort, defenders of a successful theory may simply live with what they would call an 'anomaly' to their theory, in the hope that eventually something favourable to their point of view will turn up.

The falsificationist Popper has always been fully aware of the actual uncertainties attending the falsification of theories. In his original *Logic of Scientific Discovery*, he considered the claim of an imaginary critic (whom he refers to as a conventionalist) that the demarcation criterion fails to divide theories into falsifiable and non-falsifiable ones because it is always possible to deflect the force of any falsifying instance by claiming interfering factors or problems with the observations:

I admit that my criterion of falsifiability does not lead to an unambiguous classification. Indeed, it is impossible to decide, by analysing its logical form, whether a system of statements is ... empirical in my sense; that is a refutable system. Yet this only shows that my criterion of demarcation cannot be applied immediately to a *system of statements*. ... *Only with reference to the methods applied* to a theoretical system is it all possible to decide whether we are dealing with a conventionalist or an empirical theory. The only way to avoid conventionalism is by taking a *decision*: the decision not to apply its methods.[2]

---

[2] *The Logic of Scientific Discovery* (Hutchinson, London, 1972), pp. 81–2.

So the falsifiability required by the criterion of demarcation is a matter not of logic, but of the methods used, not only in connection with probabilistic theories, but with all scientific theories. True scientists, on this view, will try to falsify their theories and will not rest content with 'anomalous' results or seek easy explanations for test failure. It is always possible to avoid falsification in these and other ways, but it is not scientific. But is it always unscientific to use evasive, 'conventionalist' stratagems in the face of counter-evidence to a favoured theory?

Kuhnian Relativism

The work of Thomas Kuhn has suggested to many that it is far from unscientific to hang on to a favoured theory in the face of counter-evidence. We have just considered the case of the deviance of Uranus and the subsequent discovery of Neptune, but the solar system has not always proved so amenable to Newtonian theory. Mercury, too, was eccentric on Newtonian principles. Encouraged by the success represented by the discovery of Neptune, scientists postulated that Mercury's orbit was being affected by an intra-Mercurial planet. A name was even assigned to the yet to be discovered heavenly body, but unfortunately for all concerned, Vulcan was not and has not been discovered. According to Kuhn's *The Structure of Scientific Revolutions*, much of normal scientific activity consists in puzzle-solving, in which 'anomalies' to theories are addressed, but without any presumption that failure to solve an anomaly, even over an indefinite span of time, will lead to the rejection of the theory. In the case of the non-discovery of Vulcan, of course, it did not, and it is easy to see why. Newtonian theory had been and was continuing to be extremely successful in many fields. It would surely have been irrational for scientists to give up a successful and knowledge-increasing model of reality just because it had come up against a minor anomaly, an anomaly that might in any case eventually be solved within the Newtonian framework. Even though Vulcan was not observed, there could be reasons why it was difficult to see, or perhaps some other factor altogether was exercising a gravitational pull on Mercury, but still within the general framework laid down by Newtonian theory.

Kuhn's view is that a theory like Newton's is rarely treated by scientists as the Popperian model suggests, as something submitted to genuine empirical testing. Such a theory is, according to Kuhn, much more like a framework within which scientists do their day-to-day work of refining observations and measurements and constructing a detailed and precise representation of the physical world. The framework itself, called by Kuhn a paradigm, is not normally up for refutation or criticism, precisely because it determines the way scientists look at the world and approach their data. The world is, as it were, questioned in Newtonian terms, and if it does not give Newtonian answers straightaway, it does not follow that it was being asked the wrong sort of question, or that a Newtonian answer will not eventually be forthcoming. As Kuhn himself puts it,

No one seriously questioned Newtonian theory because of the long-recognized discrepancies between predictions from that theory and both the speed of sound and the motion of Mercury. The first discrepancy was ultimately and quite unexpectedly solved by experiments on heat undertaken for a very different purpose; the second vanished with the general theory of relativity after a crisis it had had no role in creating.[3]

Kuhn's central analytical tool is that of a scientific paradigm. As stated above, a paradigm involves the basic theoretical framework within which scientists work at any given time, but it is rather more than a set of theoretical

[3] T. S. Kuhn, *The Structure of Scientific Revolutions* (University of Chicago Press, 1962), p. 81.

statements. A paradigm characterizes ruling sets of scientific techniques and ways of looking at data, as well as giving the very model for explanation at a given time. In view of this, it is no surprise to find Kuhn claiming that paradigms determine not simply the way scientists approach and observe data, but the data themselves. He asks whether it was mere coincidence that Western astronomers first saw sunspots, new stars, and comets moving in the supposedly immutable parts of the immutable heavens very soon after Copernicus's new paradigm had been proposed, a paradigm which undermined the previous assumptions of heavenly immutability. He also makes the point that in China, where there had been no belief in heavenly immutability, sunspots had been systematically recorded centuries before Galileo. 'The very ease and rapidity with which astronomers saw new things when looking at old objects with old instruments may make us wish to say that, after Copernicus, astronomers lived in a different world.'[4]

So far, we have been looking at what Kuhn thinks happens in science when a widely accepted paradigm rules unhindered; during a period of what he calls 'normal' science, the paradigm is protected from falsification. Counter-evidence to the paradigm, in the shape of falsified predictions, is treated as merely anomalous data, to be explained away in due course, either by discovering interfering factors which produce the anomaly, or by questioning or re-evaluating the observational evidence which appeared to go against the theory. Normal scientists will, in Popperian terminology, resort to conventionalist stratagems to deflect criticism from their paradigm. Sometimes these will produce new knowledge, as in the case of the discovery of Neptune. But sometimes they won't, as in the case of Mercury's orbit. In these cases, though, the normal scientist will attract Popperian condemnation for holding on to a theory despite counter-

[4] Kuhn, op cit., p. 116.

evidence and despite the lack of any positive explanation. Kuhn's point is that there is nothing irrational in this, nor anything unscientific; nor, in his opinion, is there any definite point at which one can finally say that a theory has too many or too difficult anomalies to make continuing to work on it irrational or unscientific.

Kuhn is here trading on the uncertainty of falsifying evidence, and he is surely right to do so. The problem is that this leaves us without a demarcation between science and non-science, and the Popperian appears to be losing at least one of the sticks with which to beat the Marxists, who, after all, devote a large part of their energies to seeking explanations for the failure of their theories. Kuhn appears to be telling us that Newtonians did the same, and never really took their paradigmatic laws to be falsifiable.

If the uncertainty over falsification were all that Kuhn had argued for, even when combined with the uncertainty over proof, there might still be a way of preserving the rationality and objectivity of science. We could cease to think of the confirmation and disconfirmation of theories as a simple matter of a single theory and its relationship to the world, and think of it rather as comparative matter, where one theory is competing with another. Mercury's orbit was tolerated as a Newtonian anomaly when there were no competitors to Newton in the offing. But when relativity theory appeared, and yielded a better prediction for Mercury, a dull anomaly became what Bacon referred to as an instance of the signpost, or a crucial experiment: a point at which competing theories could be directly compared. Before there was a competitor to Newtonian physics, it would have been irrational to give up the theory and any hope of dealing with its difficulties. Unlike the Marxists' treatment of their theory, the difficulties were admitted to be difficulties and not simply a case for reinterpretation, but they were not allowed to create a panic for the theory. However, when they can be seen as crucial between two competing theories, difficulties take on a new significance, and if all the signs point in the same direction, true scientists will take the hint and decamp to the new theory. So objectivity and rationality can be preserved, even in a world in which falsifications are not seen as occasions for the immediate overthrow of theories.

The model here is familiar and obvious. We see scientific theories in terms of their predictive consequences. Theory A has some true predictions and some false predictions, and people work on this theory, modifying it and attempting generally to show that its initially false predictions are actually explicable within the terms of the theory. Earlier observations are questioned, or it is shown that the anomalies to the theory are due to interfering factors, or the theory receives modifications of a minor sort regarding its formulae or areas of application. But some problems remain. Then theory B is put forward, which absorbs the truth content of A, by deriving the same predictions from its own theoretical perspective and, in addition, yields better predictions for at least some of the cases where A has failed to deliver true predictions. In these circumstances, we would appear to have some ground for comparing A and B, and for preferring B to A because of its better performance on the crucial instances. In such circumstances, Popper is prepared to speak of B as being 'closer to the truth' than A, even if (as will almost certainly be the case) B has anomalies of its own, either in some of A's falsity content or in areas on which A does not touch at all.

The replacement of Newtonian physics by Einsteinian can be seen in this way, with Mercury's orbit and the bending of light being among the crucial instances. And we can also see Newton's synthesis of Galileo's mechanics and Keplerian astronomy in the same way, especially if we bear in mind that Newton made more accurate predictions than Galileo on the paths of long-range projectiles and on the acceleration of falling bodies, in addition to contradicting Kepler's Third Law.

Talk of one false theory, such as Newton's, being closer to the truth than another false theory (Kepler's in this instance) seems clear enough. It has, nevertheless, come under heavy fire on technical grounds, for the notion appears to resist detailed formal exposition. But even if these problems cannot be overcome, and we have to stop describing theories as closer to or further from the truth, there would still seem to be clear rational grounds for preferring theory B to theory A, where the crucial instances appear to point to theory B. While it is open to proponents of theory A to show that this appearance may be illusory, and that the crucial instances can actually be explained perfectly well in the terms of theory A, if the instances are numerous and diverse it would clearly be reasonable to prefer B over A, at least pending a reasoned defence of A.

It is on this point that the claims of Kuhn about the nature of scientific paradigms have their most far-reaching and revolutionary effects. For Kuhn claims that the role of the paradigms involved in any piece of scientific testing make it impossible to speak of crucial instances or experiments in the intended sense. Kuhn's claim in brief is that the paradigm within which a scientist works conditions not only his theoretical world-view, but also his very observations. So when paradigm X conflicts with paradigm Y, the proponents of the one will observe things in a different way from proponents of the other. We have already seen a hint of this in the remarks about new astronomical observations after Copernicus, but Kuhn appears to take this conditioning of data by theory even further. The very phenomena which appear to one side to support their theory may well appear to the other side to support theirs. For Priestley, what we call oxygen was a confirmation of phlogiston theory, for it was (in his view) dephlogisticated air. For Lavoisier, the same phenomenon supported his own view that it was oxygen. We see the isolation of oxygen as an instance favouring the postulation of oxygen, because oxygen is part of modern chemical theory, but this is only with the benefit of hindsight. The hindsight which allows us to reject Priestley's 'dephlogisticated air' and praise Lavoisier on oxygen is one conditioned by the modern chemical paradigm. The data which we take to be crucial here are seen to point in a certain direction because we are already looking from a biased point of view.

In order to give some backing to his position here, Kuhn refers to the famous duck–rabbit of *Gestalt* psychology. The same figure and the same evidence can be used to support both the claim that what one has is a representation of a rabbit and the claim that what one has is a representation of a duck. The figure can be looked at either way, though not, to be sure, from both points of view at the same time. Nevertheless, in the *Gestalt* case, we may be able to see the lines, as a sort of observational substratum, underlying our perception of either the duck or the rabbit. This possibility of what might be called 'theory-neutral' observation is not, according to Kuhn, always available in cases of conflict between scientific paradigms, because, when competing paradigms clash, proponents of each will tend to see the evidence in their own way.

Kuhn's claim is not merely that proponents of given paradigms will miss things their opponents might easily see, as in the case of the news stars discovered after Copernicus. It is rather that the same evidence can be seen in ways that will support either point of view, depending on your initial standpoint. Thus, Kuhn has it that

since remote antiquity most people have seen one or another heavy body swinging back and forth on a string or chain until it finally comes to rest. To the Aristotelians, who believed that a heavy body is moved by its own nature from a higher position to a state of natural rest at a lower one, the swinging body was simply falling with difficulty. Constrained by the chain, it could achieve rest at its

SCIENCE AND NON-SCIENCE 71

low point only after a tortuous motion and a considerable time. Galileo, on the other hand, looking at the swinging body, saw a pendulum, a body that almost succeeded in repeating the same motion over and over again ad infinitum.[5]

In other words, what Aristotelians would see as constrained fall, Galileo saw as a pendulum. Descriptively, each perception was fairly accurate (although, as Kuhn points out, Galileo's observations of the pendulum actually had a degree of inaccuracy due to his seeing more regularity in the pendulum than was actually there). Each paradigm is thus 'supported' by the very evidence the proponent of the other paradigm would naturally claim in his own support.

According to Kuhn, then, there will be no straightforward crucial experiments or observations to provide grounds for rational comparisons of theories, at least when the theories concerned are major paradigms. If paradigms do so determine the way the data are seen, how is it then that science ever moves from one paradigm to another? Here Kuhn is forced to rely on psychological and sociological explanations. Periods of 'normal' science—the state in which scientists work calmly within a particular paradigm—are occasionally interrupted by periods of revolutionary turmoil. These may occur because the number and extent of anomalous instances seem to be on the increase, and individual anomalies increasingly resistant to solution. In such circumstances, scientists will be open to new points of view, and some will develop them. But if and when the majority of scientists and the scientific community as a whole adopt a new paradigm, it will not be for reasons which are justifiable from a neutral point of view. After emphasizing the way in which defenders of an old paradigm can always quite reasonably continue to defend their point of view (and often do until they die off), Kuhn speaks of the basis,

though it need be neither rational nor ultimately correct, for faith in

5 Kuhn, op cit., pp. 117–18.

the particular candidate chosen. Something must make at least a few scientists feel that the new proposal is on the right track, and sometimes it is only personal and inarticulate aesthetic considerations that can do that. Men have been converted by them at times when most of the articulable technical arguments pointed the other way. When first introduced, neither Copernicus' astronomical theory nor De Broglie's theory of matter had many other significant grounds of appeal. Even today Einstein's general theory attracts men principally on aesthetic grounds . . .[6]

To the individual motivations of individual scientists, Kuhn and his followers would add the social effect of authority in the scientific community: the way publication, preferment, and research money will be distributed by those at the top of the community and in accordance with their favoured paradigm. In brief, then, a new paradigm takes over when it captures the hearts and minds of the commanding figures in the scientific community of the time, and this process is neither wholly rational, nor perhaps, according to the ideal of the open scientific community, wholly admirable.

There is more than a dash of relativism in all this, which Kuhn does little to mitigate in his frequent underlining of the re-emergence in later paradigms of assumptions once held in scientific discredit. Thus, for example, the notion of action at a distance, basic to the medieval and Aristotelian world-view, had seemed frightfully unscientific to the mechanists of the seventeenth century, but it was (reluctantly) adopted by Newton in his theory of gravitation. The notion of local action, implicit in current talk of forces and fields, would also have seemed a throw-back to more mystical ways of thinking to seventeenth-century mechanists, and perhaps to Newtonians too. Similarly the Aristotelian notion of objects having intrinsic powers and propensities can be seen as having been revived in modern physics.

I shall argue in Chapter 6 that some of what Kuhn says

[6] Kuhn, op cit., p. 157.

about the unfalsifiability of scientific paradigms, particularly at the level of their core or 'metaphysical' assumptions, is correct. But one could still maintain this without going all the way with Kuhnian relativism on scientific theory. One could argue that, at a given time and in a given formulation, one paradigm was better than another, in the sense of accounting better for the evidence. To say this, though, one would have to be able to see the evidence as providing a neutral testing ground, on which one could come to a reasoned evaluation of the theories. But it is the existence of such a neutral testing ground that Kuhn wishes to deny. And this is what makes his position relativistic, for, on his account, there can on occasion be no good reason for preferring one paradigm to another, even when all the scientists involved in the discussion see the same features as desirable in a good scientific theory: accuracy, scope, simplicity, fruitfulness, and so on.

According to Kuhn, though, the absence of a shared observational perspective on the data means that we can no longer say that scientists operating from different paradigms see the same thing. There is no access to the world and no classification of objects independent of human activity and conceptualization. We have already seen that this is a consequence of Goodman's new riddle of induction. Classifications and groupings of things depend on points of view. This thought is blown up by Kuhn, and even more by Paul Feyerabend, into the further thesis that people classifying from different points of view may be unable to find any neutral ground on which to communicate. Their thought-patterns are not mutually translatable, and their theories are not strictly comparable because they are talking about different things. And in science, according to Kuhn, this difficulty arises with a vengeance because in science the basic similarity–dissimilarity relationships, on which observation and classification depend, are themselves dependent on the very paradigm within which we work.

Under a paradigm change, 'objects which were grouped in the same set before are grouped in different sets afterwards and *vice versa*. Think of the sun, moon, Mars, and earth before and after Copernicus; of free fall, pendular and planetary motion before and after Galileo; or of salts, alloys, and a sulphur–iron filing mix before and after Dalton.' Most objects continue to be grouped in the same sets as before, but the changes crucially affect the network of interrelations among sets: 'Transferring the metals from the set of compounds to the set of elements was part of a new theory of combustion, of acidity, and of the difference between physical and chemical combustion. In short order, those changes had spread through all of chemistry.' And this is where we come to the key point: 'When such a redistribution of objects among similarity sets occurs, two men whose discourse had proceeded for some time with apparently full understanding may suddenly find themselves responding to the same stimulus with incompatible descriptions or generalizations.' And then, presumably because of the way the paradigm involved suffuses the way each sees the world, neither can explain in mutually intelligible terms to the other his use of the term 'element' or 'mixture' or 'planet' or 'unconstrained motion'. Because of this 'the source of the breakdown in their communication may be extraordinarily difficult to isolate and bypass'.[7]

The full Kuhnian thesis, then, is not the ordinary first-level relativism of having no rational way of deciding between competing paradigms. It is actually the more dramatic claim that there is a 'breakdown of communication' between defenders of 'incommensurable' paradigms, who because of their paradigm-led observations may be said to see the world differently, and even to live in different worlds. This may be a relativism beyond straightforward disagree-

[7] T. S. Kuhn, 'Reflections on My Critics', in I. Lakatos and A. Musgrave (eds.), *Criticism and the Growth of Knowledge* (Cambridge University Press, 1970), pp. 231–77, at pp. 275–6.

ment in that, if taken to its logical conclusion, such radical incommensurability would, because of a lack of neutral translation, make it impossible to know whether there was real disagreement between proponents of competing paradigms. But the effect of this view is certainly relativistic. It makes the history of Western science look like a rationally unjustifiable series of lurches from one closed theoretical and perceptual framework to another, with no possibility of mutual communication or evaluation.

Kuhn derives his thesis of incommensurability philosophically from certain theses concerning the lack of cut-and-dried criteria for the testing of scientific theories and the need to presuppose standards of similarity and dissimilarity in any observation one makes of the world. But he fleshes out these theses with appeals to the history of science. And this, in turn, raises questions both as to the adequacy of his history and, more generally, as to the nature of the relationship between the history and philosophy of science.

The Relationship between the History and Philosophy of Science

We shall return in the next chapter to a straightforward assessment of Kuhnian relativism. Before this, it is worth devoting a little time to a question we have so far sidestepped, despite the fact that it has been present only a little below the surface of much of what we have been considering. Right at the start of the book, I spoke of growth of knowledge in science as being an unquestionable fact. This at once put a certain perspective on our philosophical reflections on science, and it is a perspective which comes from our having some rudimentary sense of a certain passage in Western intellectual and social history. Right from the start, our philosophy of science is to be seen as a philosophical reflection on certain historical events, bounded and shaped by our sense of what that history is. Had there been no 'rise of modern science', our philosophy of science would doubtless have been very different. The owl of Minerva flies only at dusk, as Hegel observed: philosophy is always a reflection on, and hence after and necessarily after the events.

I have presented both induction and falsification—as practised in the hands of Bacon and Popper—as philosophies of the rise of modern science, as attempting, in other words, to explain what that rise consisted in and, to an extent, how it was achieved. But now, on the stage, walks Kuhn, a historian first, telling us that this view of the rise of modern science is all wrong. There was no steady progression of knowledge, but a series of revolutions, of incommensurable changes of world-view: mechanism to Newton, Newton to Einstein, Priestley to Lavoisier, Aristotle to Descartes, and perhaps even back again in contemporary sub-atomic physics. Maybe, in addition to discontinuity, there is even no clear growth of knowledge. Those guided by the mechanist paradigm, treating all physical processes as clockwork interactions, scoffed at Aristotelian forces and propensities, but in doing so did they not lose an insight we are only beginning to regain into the immaterial, non-corpuscular aspects of material things and interactions?

Intellectual life is, like everything else, subject to specialisms and the division of labour. Philosophers of science specialize in science and its problems, and—to their loss—may know little of history, and less of its methodological problems. To put it bluntly, had philosophers of science known more history and thought more about the philosophy of history, Kuhn's account of scientific *revolutions* might not have been received with the awestruck horror or enthusiasm it has tended to evoke. People might have asked more about Kuhn's historiography, and jumped less quickly to man the barricades, on one side or the other. We can all recall those tiresome questions in history exams, about whether a particular so-called revolution was really a clean break from the past, or whether it was simply the formal enactment of changes in society and political structures which had been long fermenting. And yet, when Kuhn invites us to see the history of modern science as a series of sharply discontinuous revolutions many of us forget all we know of political and social history and its methods.

In fact, as Peter Munz has put it,

anybody even superficially acquainted with the study of the French Revolution and the vast literature on the subject will immediately recognise the weakness of Kuhn's contention ... It is perfectly possible to tell the story of the French Revolution either as an addition to the *ancien régime* or as the story of the violent and revolutionary overthrow of the *ancien régime*. It depends entirely on the series of events the authors have selected.[8]

And Munz gives the examples of Burke and de Tocqueville who have taken opposing views of the French Revolution; historical knowledge is not something which can simply be read off from the 'facts'. As much as any other enquirer, the historian must select and emphasize and interpret his material. Interpretation becomes particularly significant when the historian is trying to establish theses about the causes and origins of events, and about the significance of such things as revolutions.

The role of interpretation is crucial when one comes to write a history of science. Kuhn would no doubt criticize histories of science which assumed that science involved a steady growth from less to more knowledge, or which assumed that science went from one falsified theory to another, as soon as a falsification appeared. But, one might ask, how else would an inductivist or a falsificationist write a history of science? The nature of science is, as we are seeing, a philosophically contentious issue. The historian of science has initially to approach his material with some sense of what

---

[8] *Our Knowledge of the Growth of Knowledge* (Routledge & Kegan Paul, London, 1985), p. 121.

science is and what scientific progress is like, in order to select the *scientifically* significant pieces of evidence from the behaviour of the people we call scientists. There is, in the history of science, always an element of rational reconstruction of untidy events, in the sense that no scientific event is ever entirely pure, methodologically speaking. The historian of science will be interested in the development of scientific knowledge, and there will always be some distinction to be drawn between what bears on that and what belongs rather to the scientifically extraneous features of the situation: the politics, sociology, and psychology of the scientists, and the rest.

So, if you see science in Baconian terms, you will naturally highlight the collecting and analysing of data involved in any theory-building, and show how imaginative leaps, if any, were tightly controlled by the data. If you are a falsificationist, you will emphasize the testing of theories and praise scientists as being 'truly' scientific when they discard falsified theories. And if you are a Kuhnian, you will tend to focus on discontinuity and irrational elements in the history of science.

Does an initial philosophical standpoint which guides one's emphases and choice of data in writing a history of science mean that one can never get nearer the truth in the history of science? This would certainly be a depressing conclusion, and one which would at the same time limit the rational development and examination of the philosophy of science. But there is no need to draw this conclusion. Study of the history of science can show us that a given philosophy of science is at best a partial account of what goes on in science. I tried to reveal this through my criticisms of Bacon: a Baconian account would inevitably gloss over and distort much of significance in the work of Kepler and Newton. But, against Popper and Popperians, there are things right in inductivism, and this too emerges in considering the history of science. The relationship between the history and philosophy of science ought to be one of checks and balances, a philosophy inspiring the initial foray into history, but always being open to historical counter-evidence by which the philosophy itself may eventually come to be modified if it comes to seem a travesty of what actually happens in 'good' science, where there is growth of knowledge. It can seem a travesty if one continually and systematically overlooks aspects of scientific practice, by underplaying the significance of evidential support in scientific practice, or by downplaying the actual degree of effective mutual communication between proponents of different paradigms, or by overlooking the way scientists actually manœuvre to deflect the force of counter-evidence. While the writing of a history of science will, for the reasons given, be guided by the historian's initial philosophy of science, this does not mean that all the vast mass of material available to the historian of science is so guided, or that other reconstructions of that material might not give a more coherent account of the material than the initial philosophical standpoint, or, indeed, that the historian might not come to modify this standpoint by a realization that the historical material might be organized more coherently by a different philosophy.

But the realization that any history of science will have to start from a certain philosophical position should make us cautious. We should be wary of histories which appear to 'prove' in a couple of hundred pages the very philosophy which inspired them, and should be on the look-out for other histories which might suggest that things have been overlooked or over-simplified in the initial history. And in the case of Kuhn, we should not be surprised to discover that many historians of science stand as the Tocqueville to his Burke, stressing and highlighting continuities Kuhn overlooks. I will argue in the next chapter that these historians have a valid point, even though, as I will suggest, there is a grain or two of truth in Kuhn in his attacks on inductivists and falsificationists. In Kuhnian terms: we should not forget that Kuhn himself, in writing his history, is operating with his own paradigm, which seems at times over-close to the way he selects and interprets his data, which he then uses in circular manner to support his paradigm. But rather than simply saying that our stance pro- or contra-Kuhn, as the case may be, is simply a matter of irrational paradigm choice on our part (as on Kuhn's), perhaps we should be ready to look again at the history of science to see if a Kuhnian account is the only or the best interpretation of the data.

Without anticipating too much of what will be said in the next chapter, it is worth underlining at this point the basis on which Kuhn's relativistic position actually rests. There are, as we have just seen, reasons to be suspicious of Kuhn's historical data, both because the data are selected and analysed too much in the light of Kuhn's relativistic philosophy of science and because (as other historians such as Shapere have argued) it is far from clear that the development of Western science is as disjointed as Kuhn suggests. But, in a way, all these appeals to and arguments about history are irrelevant. Let us suppose that what Kuhn says about history is correct, and that the development of science is thoroughly discontinuous and fragmented. That by itself would do nothing to establish any relativistic conclusions. It would have no tendency to show that reasoned judgements could not be made about the superiority or inferiority of one set or other of theories. And what Kuhn uses to establish that there can be no such reasoned judgements is simply the familiar claim that there is no cognition of the world independent of and apart from the concepts we use to approach the world, together with the further thesis that each set of human enquirers is imprisoned within their own conceptual scheme, unable to communicate with or enter into mutual comparisons with those belonging to other conceptual schemes. But this is to say much more than that the history of science is disjointed and revolutionary. It is to make the fundamentally philosophical claim that something in the nature of our conceptualizations of the world makes it impossible for observers from different theoretical backgrounds ever correctly to recognize that they are talking about or observing the same things. In the next chapter, I will argue that nothing of the sort follows from the mere fact that all our knowledge of the world is mediated through our conceptualizations and categorizations.

5

Observation and Theory

Observational Common Ground between Theories

Central to Kuhn's paradigm-led philosophy of science and to much other work in contemporary philosophy of science is a denial of any real or sustainable distinction between observation and theory in science. Both inductivist and falsificationist philosophies assume a working distinction between theory and observation. Observations support or refute theories precisely because they are in a significant way separate from them and can be regarded as independent of the theories they stand in an evidential relationship to. There would be an obvious circularity if a theory was supported by a piece of evidence which could be accepted in its intended interpretation only if the theory under test was true. Equally, it would be no refutation of a theory if the evidence supposedly refuting it had been selected and interpreted on the supposition that the theory in question was false. Yet such claims are inherent in Kuhn's view of paradigm comparison, and they make neutral comparisons between theories impossible.

In considering the relationship between theory and observation in science, it is important to distinguish a weak thesis about the suffusing of observation by theory from a much stronger one. The weak thesis says that all observations are conditioned by presuppositions, assumptions regarding similarity and dissimilarity, directions of interest, and so on. Though often referred to as the thesis that all data are 'theory-laden', this thesis amounts to little more than the positive side of the criticisms we made of Baconian hopes for presuppositionless observation. The fact that there are interests and schemes of classification behind any observation of the world does not amount to an elevated sense of 'theory'. It need not imply that precisely formulated or systematic assumptions are guiding one's observations, and is quite consistent with pretty random and undirected noticings of aspects and features of one's environment. Some of the features one notices by chance may indeed be quite hard to reconcile with one's explicit theories about the world, and pave the way for revision of those theories; to that extent one can think of a lot of one's observations as, in a significant sense, pre-theoretical or non-theoretical. Whether or not one is willing to do that, however, it is crucial to see that the weak thesis about the theory-ladenness of all data does not entail the strong thesis.

The strong thesis is that the observing of data in science depends crucially on the paradigm one holds, that one's observations are always biased in favour of that paradigm. We have already seen that this is Kuhn's view. Aristotelians saw constrained fall, where Galileo saw a pendulum; what Priestley took to be dephlogisticated air, Lavoisier sees as oxygen. It is this stronger thesis which Kuhn needs in order to establish incommensurability between paradigms in particular and his relativistic thesis in general. If we could establish a level of observation regarded by all parties to a paradigm dispute as neutral between their respective paradigms, comparisons could be effected through assessing the ability of each paradigm to deal with the data agreed by all to be the data to be explained. Not surprisingly, it is just this type of assessment which Kuhn is at pains to deny by claiming that, in a transition from one paradigm to another, there will normally be no more primitive observational level, common to both parties, other than their theory-dominated level of observation. This is the significance of the talk of Aristotelians *seeing* constrained fall and Galileo *seeing* a pendulum, and Kuhn is right to comment that at that level of discourse, the two parties to the dispute are not seeing the same things. Pendula are not stones in constrained fall, nor is oxygen dephlogisticated air. The pairs of concepts may not match up even extensionally; different things may well fall under the concept. Applying one member of the pair rather than the other will then lead scientists to make quite different measurements and further observations. And, in Kuhn's view, such things as oxygen and pendula (and, presumably, dephlogisticated air and constrained fall) will be 'the fundamental ingredients of [the] immediate experience' of scientists.[1] So the hope for a neutral, more fundamental, and less theoretically conditioned level of observation is apparently doomed, and our paradigms remain incommensurable, hermetically sealing in those inside and sealing out those without.

But is this so? A quick way to deal with the incommensurability position, as applied to the history of Western natural science at any rate, would be to say that Kuhn himself, in his treatment of pendula and oxygen, does just what the thesis supposes one is unable to do. He describes the content of various paradigms in a way which presupposes the truth of none of the paradigms concerned. And he also describes the phenomenon to be explained in a way which does not presuppose any particular explanation of it. In the case of the pendulum, this is quite explicit in the words with which he introduces the topic: 'since remote antiquity most people have seen one or another heavy body swinging back and forth on a string or chain until it finally comes to rest'. 'Heavy body swinging back and forth on a string or chain' may be a theoretically loaded description, but not in the sense that the truth of either Aristotelian or Galilean explanatory dynamics is presupposed.

If we are interested in explaining and predicting this

[1] Kuhn, *The Structure of Scientific Revolutions* (University of Chicago Press, 1962), p. 127.

phenomenon at that relatively neutral level of description, we may well come to see that the Galilean account is preferable in terms of simplicity and in terms of fruitfulness. As Kuhn himself points out, on the Aristotelian account, the motion of a pendulum was an extraordinarily complex phenomenon to describe and explain; also, the Galilean account led to the discovery of many other properties of the pendulum, and subsequently to much else besides. Even if both accounts were empirically equally adequate regarding the behaviour of the pendulum, in these circumstances we would clearly have good reasons, related to the internal goals of science, for preferring the Galilean paradigm.

There would be no need here for an irrationally motivated decision in favour of the one, and against the other. And one can maintain this position even while admitting that the temporarily defeated position might ultimately be correct. With science, as in a court of law, a decision can only be made on the evidence available at the time of making the decision, and the most reasonable decision relative to a body of evidence may be wrong at the end of the day. It is always open to those who have a hunch in favour of a temporarily defeated paradigm to look for further reasons in its favour, even while admitting reason is for a time against them. But Kuhn's position is not this sensible and harmless fallibilism: it is rather a denial that reasoned decisions of any sort can be made between paradigms.

I said that our appeal to Kuhn's own text with its theoretically neutral talk of heavy bodies swinging back and forth was a quick way with the incommensurability point. Without wishing to downplay either the force or the validity of this, it is possible to go somewhat deeper in criticizing Kuhn. The point we have made so far is that there may often be observational common ground between competing paradigms, on the basis of which reasoned comparisons of the paradigms may be made. But let us concede to Kuhn that there may on occasion be no handy neutral description of any observational common ground and that the very discovery of what might later be seen as the observational evidence supporting one paradigm and falsifying the other is only made possible by the new paradigm. Such may often be the case in theoretical science. It is possible to maintain, even in these circumstances, that the incommensurability thesis is false, and that reasoned comparisons of the competing paradigms can be made in virtue of the observational ground which stands between them.

The type of case which might be thought to favour Kuhn here would be when a type of entity not previously identified is postulated by scientists. In such circumstances, it is highly likely that the terms in which this is thought of are highly 'theory-laden' in the full-blooded sense. The postulation and identification of the new type of entity brings with it a whole lot of explicitly theoretical assumptions, which may be deeply embedded within one of the paradigms in question.

Mary Hesse has suggested that we look at the discovery of Jupiter's moons in this light. At the time of their discovery by Galileo, many highly contestable and theoretical assumptions were involved, about the nature of moons and planets, and indeed about the solar system and the heavens generally. Galileo's observations would seem to be a classic instance of theory-laden observation. But further work on the subject has exorcised their theory-ladenness in such a way that they are now accepted as common-sense objects, identifiable and reidentifiable, regularly orbiting Jupiter, and even visitable by space probes. The initial theoretical background involved in their discovery has largely dropped away, and the objects concerned can be recognized as ordinary and commonsensical as, say, the Galapagos islands. This observational ground which might now allow us to prefer Galileo to Aristotle without a second thought has become common ground in and through the theories espoused by Galileo, but it is now free of those theories and independent of them. Even an Aristotelian, if rational, would now have to admit existence, as will any future astronomical theory, however few of the theoretical assumptions of Galileo or his immediate successors it takes on board.

Let us follow Hesse in contrasting the case of Jupiter's moons with those of phlogiston and electrons. Phlogiston, too, was a highly theoretical notion. Many properties were ascribed to this supposed indirectly observable gaseous fluid, and in contrast to the moons of Jupiter we now say that there is no such thing. Phlogiston theory, though, undoubtedly drew people's attention to many of the effects which were ascribed to phlogiston, for example, that a flammable gas was emitted when sulphuric acid was poured on zinc. Phlogiston theory may well have inspired Priestley to make this experiment, and he thought that through it he was observing phlogiston. We now say that what was given off then was hydrogen, and that there was no single stuff which has all the properties ascribed to phlogiston. The stuff identified as phlogiston did not, after all, constitute a single natural kind, with a common underlying structure. But it is certainly the case that chemists after Priestley had to account for what he took to be the effects of phlogiston, including the giving off of hydrogen. Relevant observational evidence may emerge only through the operation of a paradigm, but in such a way that it survives the rejection of the paradigm, and even comes to form part of the reason for that rejection. Something like this happened in the phlogiston case, precisely in the realization that the various effects ascribed to phlogiston were actually produced by different substances.

The electron example poses the same issues as the phlogiston case, but highlights them in a different way for we do not as yet know the outcome of the case. As Hesse remarks in her paper, we know a fair amount about certain sub-atomic objects which have an observable charge and mass, and which we can distinguish from other micro-particles in a number of experiments. In this sense, we know a lot 'about' electrons. Do electrons exist, then? Hesse argues that in view of the volatile state of contemporary micro-physics, we cannot rule out a scientific revolution in which electrons—at least as currently understood—are held not to exist. Much of what we believe now about electrons may come to seem false both in detail and in general, as happened with phlogiston. But the local and particular truths we now interpret as being about electrons will remain. Just as phlogiston/hydrogen goes up in flames, so 'electrons by any other name activate cathode ray tubes and geiger counters'.[2] As with phlogiston and hydrogen, there will be no strict commensurability between electrons and the new particle or particles or whatever it is the new theory postulates as the cause of cathode ray tube and geiger counter activity. Before and after some revolutions there may in a sense be different things observed; the observations are conceptualized quite differently. But some at least of the particular effects of the things conceptualized differently will be the same, and there is thus a significant degree of continuity of scientific knowledge through even major conceptual revisions. At the grand theoretical level, one can concede to Kuhn that precise mappings of theoretical terms and concepts may present formidable problems. But science also works at a less elevated level, recording and discovering effects and regularities of a more mundane sort. These are not lost in conceptual revolutions, even where a discarded conceptualization prompted their discovery. Their existence and persistence through revolution provides the observational, non-theoretical basis on which the theoretical superstructures of science rest, and by means of which the theoretical superstructures can be compared for such things as predictive accuracy, scope, simplicity, and fruitfulness.

    [2] Mary Hesse, 'Science Beyond Realism and Relativism', forthcoming in the Proceedings of the Conference on Cognitive Relativism and Social Science, Utrecht, 1986.

Observation and Theory

If anything has become a received idea in recent philosophy of science, it is the thesis that there is no sharp distinction in science between observation and theory; in other words, that there is no pure observational level in science which stands free of theoretical baggage. Some of the reasons for this view are good and some not so good. In this section, I want to suggest in a non-partisan way that there may be something to be said for the distinction when it is properly understood, and that, properly understood, it is an important distinction for understanding the nature of science.

The reason the observation–theory distinction initially fell into disrepute was to do with the failure of the logical positivists of the 1930s to develop a pure observation language, which would enable objects and events in the world to be described in a way which made no unverifiable assumptions about what was being observed. The hope was that if a purely observational part of language could be isolated, or even constructed, science would rest on firm foundations. Unfortunately for this project, the observational language we use in everyday circumstances came to be regarded as being thoroughly theory-laden. Talk of cats, dogs, tables, chairs, trees, people, and so on, would bring with it all sorts of assumptions about the nature and future behaviour of the things being observed. Many of these assumptions were highly unverifiable from a logical positivist perspective, as they embodied expectations regarding future states of the objects concerned, uncheckable beliefs about their origins, and, most problematic of all, hypothetical or counter-factual beliefs about how they would have behaved in circumstances which never actually occurred. My belief that I am currently writing on a table carries with it the implication that if some other normally sighted observer had come into the room ten seconds ago, he would have been able to see the table. But as no such observer did come in, this is not an implication which can ever be confirmed or checked by direct observation.

On the other hand, restricting the language of observation to one's present states of consciousness, which would avoid the difficulties of verifying ordinary observations, was quite inadequate for the purposes of science. In science, falsification and confirmation is through objective and inter-subjectively observable states of affairs, not through subjective reports of observers' momentary states of mind. But there seemed to be no way of describing objective states of affairs which was assumptionless, even at the everyday level. In any case, many of the observations relevant to scientific testing are, as we have seen, highly dependent on theory. Calculating the mass or orbit of a planet is no straightforward matter. Even observing it through a telescope involves the tacit acceptance of large tracts of optical theory. And if we are to regard telescopes as theoretically contaminated, as it were, what of our own sensory faculties? To regard them as pure of all assumptions and presuppositions is to overlook the way in which points of view, unarticulated classificatory assumptions, and focuses of interest are embodied in our sense-organs, which are attuned more to certain aspects of the physical world than to others.

The upshot of all this was to convince many philosophers that there was no viable distinction between observation and theory, and little point in trying to draw the distinction anyway. Even at the most primitive level, it seemed, observations were suffused with presuppositions and assumptions. Some observations, like those of planets or of electrons, might be theoretical, relative to our unaided perception of the physical world, but it is all very much a matter of perspective and degree. We are always immersed in theory and presupposition, and attempts to draw a clear-cut distinction between observation and theory serves only to disguise this fact and to fail to appreciate its significance.

That all observation involves presuppositions and assumptions is undoubtedly true, but this does not imply that there is no point in drawing distinctions between more and less theoretical levels of observation. Philosophers of science who see everything, in science and everyday life, in terms of theories may be unable to see any point in doing this. For such philosophers, as the intellectual successors of logical positivism, once the presuppositions involved in everyday observational talk became apparent, all becomes theory, from the most banal talk about glasses of water to the most recherché trackings of mesons and muons. There is no essential difference between my statement, 'Here is a glass of water', and J. J. Thomson's first faltering measurements of the mass of the electron.

In *The Logic of Scientific Discovery*, Karl Popper actually uses the statement 'Here is a glass of water' in order to demonstrate the unverifiability and theory-ladenness of even the simplest description: it is theoretical because the words 'glass' and 'water' denote physical bodies which exhibit law-like behaviour in the future as well as in the past, and unverifiable because this is something we cannot know when dealing with a particular object and the liquid it contains.[3] The upshot is that we simply decide to accept certain observation statements for the purposes of science and everyday life, but, according to Popper, we can give no justification for this. Absolutely speaking, glasses of water and electrons are on a par, and our decision to take the one rather than the other as our observational basis is a psychological fact, something about us, but not one that gives our talk of glasses and water any firm grounding in reason.

Even if we accept the intuition underlying this approach to observation, some significant differences between observations and theories are still being overlooked. As we saw in the previous section, there may well be truths bound up in the observation of particular instances of moons, hydrogen, and

---

[3] Cf. *The Logic of Scientific Discovery*, p. 95.

electrons which are separable in principle and in practice from whatever theories are current about these things. And the same goes for water. To say, with Popper, that a belief that a certain liquid is water may bring with it certain expectations that it will manifest certain regular or law-like behaviour, is not at all the same as holding any explicit theory about water. We might also be able to check the current behaviour of a liquid sufficiently to ascertain that *it* is indeed water now, and, this again is different from verifying an explicit theory about its future behaviour, let alone a theory about the chemical composition of all instances of water throughout the whole of space and time. These differences between observation statements and theories are important because they form the basis of our ability to see an observational common ground surviving theoretical change, and they can be lost sight of if we accept uncritically the doctrine of the 'theory-ladenness' of all observations.

At the same time, proponents of the theory-ladenness of observation are right to emphasize the significance of presupposition, and sometimes of explicit theory, in directing and forming our observations. There may then seem to be little to distinguish the observing of a glass of water from that of a planet or even of an electron. Both involve a highly selective interpretation of specific features of the environment, and, from an absolute point of view, it may be hard to see why one observation should count as more basic than either of the others. Even the fact that instruments have to be used by us in one case, but not in the others, or that one observation is indirect where the others are direct, may be of little significance when we consider that the very sense-organs which yield the supposedly direct observations are themselves highly refined and selective instruments for observing the world, and highly fallible to boot.

Absolutely, there may be nothing to choose between observations of electrons and of glasses of water, but we, as human beings, do not stand in an absolute relation to the world, with every aspect of the world equally open to us. As human beings, we are naturally attuned to certain features and aspects of the world; attuned, that is to say, by our sense-organs and intellect, to survival in and interaction with the world at a certain ecological level. So there is a very good reason for us to take glasses of water as more basic and less theoretical than electrons, or even than moons and planets. Observing and interacting with such things as glasses of water is what we are, as it were, programmed by nature to do. By contrast, forays into the very small and the very big are thus inevitably going to be indirect, and less than immediate. We are not naturally attuned to such worlds, and we will need instruments to aid us in investigating them.

Our place in a certain niche of existence gives some point to a distinction between more and less theoretical levels of observation, and suggests why we do not just dogmatically 'decide' to accept some levels of observation as basic. We accept as basic those which relate to our lived experience of and interaction in the world, and this ramifies out into our sense that the theories of science, in so far as they are acceptable, will frequently have practical effects in the world of experience. And this basic level of observation, being related to our genetic inheritance, and our needs and interests as human beings, provides a common ground for communication between people from different cultural backgrounds. This sharing of sensory apparatus and of interests, makes it most unlikely that human beings even from the most widely different cultures would be completely unable to communicate at the level of the observations relevant to basic survival. Even less, then, is it likely that there should be a Kuhnian breakdown of communication between people—such as modern Western scientists—engaged in recognizably the same enterprise.

Empiricism

Our sharing of a lived world, of needs in that world, and of a sensory apparatus somewhat adapted to that world, is part of what being human consists in. Our common humanity and our common world make communication possible across theoretical and cultural divides. This will be as true in science as anywhere else, and should make us reluctant to accept too readily 'breakdowns' of communication between different theories and cultures. In fact, as Donald Davidson has argued frequently,[4] understanding and translating the utterances of others requires that we share with them a good measure of beliefs about a common world and interests in that world. Making sense of the words and actions of others is a process of crediting them with *intelligible* beliefs and desires: beliefs and desires, that is, which seem in the main to be consistent in themselves and to reflect what we can see as credible attitudes to the common world in which we all live. Attributing obviously false or unintelligible beliefs to others leaves us with the problem of explaining why they should come to have them. When the bizarre beliefs we attribute to others are ones that they ought not to have, given the evidence that stares them in the face and the logical implications of other beliefs we credit them with, we may have to question our interpretation of the concepts we are attributing to them, or fail to make sense of their words and behaviour at all. Some beliefs may appear, under a given interpretation, so bizarre as to make us uncertain that the words we are translating really do refer to what we are taking them to refer to. This point will apply particularly when we are dealing with words for objects used in everyday activity. Suppose we think a native word refers to chickens. Our confidence that our translation is correct will decline if the natives sometimes fail to assent to it when to us it is obvious that there is a

[4] Cf. his *Inquiries into Truth and Interpretation* (Clarendon Press, Oxford, 1984), especially Essays 9, 10, 11, and 13.

chicken before us, or if they deny that chickens have two legs or come from eggs. In other words, attributing referential terms to others involves a basic measure of agreement between us and them about just what it is we are referring to, and about some basic characteristics of the object referred to.

We can, though, tolerate a good deal of diversity in belief at the theoretical level. Some of us talk about electrons, no doubt, and some of the Azande poison chickens to establish marital misdemeanours. But both we and the Azande can recognize chickens and share a good deal of humdrum knowledge about the types of creature they are, and the same goes for cattle and children and huts and rivers. This does not mean we share all the same beliefs, any more than Priestley and Lavoisier shared all beliefs about hydrogen (or phlogiston). In fact, the two types of case are not dissimilar. For the Azande, but not for us, a chicken may be an oracle, and for us (or some of us), but not for the Azande, it occupies a certain place in evolutionary taxonomy. But none of this prevents us and them from sharing a whole lot of humdrum beliefs *about* chickens, beliefs which form the basis for us thinking that we and they are talking about the same things when we disagree about their natural history or their oracular powers.

The 'humdrum' beliefs we and the Azande share about chickens and much else are, of course, those beliefs which stem most directly from our basic human perceptions of and activities in the world to which we are adapted. Although through our theories, and the instrument-aided observations they lead to, we can go beyond and correct some of the pre-theoretical picture of the world we have by virtue of our being human, there is always going to be a sense in which all our knowledge and theory is based on elements in that picture. For it is in our statements about chickens and the like that our most basic interactions with the world are expressed. More theoretical knowledge of the world is always going to have some connection, however remote, with the humdrum level if it is to count as science fact rather than science fiction. Reflection on our evolutionary development, as adapted to and with faculties adapted to a particular aspect of the world, can suggest to us why this is so.

Empiricism is the view that our knowledge of the world is anchored in our sensory interactions with the world. What I am suggesting here is that there are certain aspects of the world to which our sense-organs are naturally attuned, and that these aspects form the basis of our everyday, humdrum picture of the world, as populated with relatively stable and enduring objects of medium size. Not everything in this picture is thereby true or accurate, but it inevitably forms our epistemological starting-point for it is where we are sensorily most responsive to the world. We notice things at this level and respond to them most naturally. Our experience tracks how things are most sensitively at the level of medium-sized dry goods; that is to say, at this level of the world, changes in the world are most readily registered by changes in our experience. We naturally and correctly expect observationally more remote theories to make some contact with the world of everyday experience, even if only at the level of registering meter-stick readings and traces on screens. This is because, in the end, theory has to answer to our pre-theoretical, everyday observations and experience. If it does not, it loses contact with the data it is ultimately a theory of, and in terms of which it proves its acceptability.

Having said all this about the empirical basis of any acceptable theory in science, and the roots of our sensory faculties in our evolution, the fact remains that a great deal of current scientific knowledge is not of readily observable things or processes. Much, indeed, is about entities and powers, which even on a generous interpretation of observation, are actually unobservable. It thus becomes a pressing question to investigate the relationship between the observable and the unobservable in scientific theory.

Unobservability and Underdetermination of Theory by Data

Having established in the earlier sections of this chapter that there is some point in basing our science in what is primitively observable for us as human beings, I now want to sketch a problem which arises for our philosophical understanding of science when a theory of science attempts to go beyond the observable. There are two different ways in which theories of science may do this. They may simply postulate the existence of types of entity which our current instruments and powers of observation do not give us direct observational access to. Leaving aside temporarily any doubts about observation at the sub-atomic level of reality, as a neat example of the way something initially unobservable becomes observable, we may take H. Yukawa's postulation of a new type of particle in 1935, partly to explain the bonding of atoms. A new generation of accelerators experimentally confirmed the existence of such a thing, known subsequently as the meson, in 1947. On the other hand, we also have in science cases where, for one reason or another, observation is never going to reveal directly the nature or the existence of the entities or forces we postulate in order to explain our data. As we shall see, Berkeley saw Newtonian forces in just this light: at most their effects could ever be observed. If Berkeley is right, there is here a radical and permanent underdetermination of theory by data (and this led him to scepticism about the truth-claims of the theories concerned). What I want to consider in this section is the implication, for our understanding of scientific theory, of the existence of the underdetermination of theory by observational data, whether this underdetermination is temporary—and our data later increase so as to make what was once merely theoretical observable—or permanent—as when no imaginable increase of our observational powers could lead us to observe directly the things our theories invoke to explain the things we can observe.

Many theories of science do attempt in one or other of the ways we have considered to go beyond the observable. In order to explain the observable, they speak of entities and forces which cannot be observed, but only theorized about. Wherever one opts to fix the boundary between what is observable and what is unobservable, problems arise once one attempts—as science characteristically does—to explain the observable by the unobservable. The boundary-fixing question is one which arouses a great deal of passion; it is perhaps better to side-step it here and return to it in the next chapter. I will begin by illustrating, with a deliberately simplistic example, the predicament which arises whenever in science we attempt to explain the observable by invoking the unobservable. With what I am calling the predicament in mind, we can then consider some real-life candidates for unobservable entity status.

The example I want to consider is due to Hilary Putnam.[5] We are asked to consider a world consisting simply of a straight line. Now, the geometry of straight lines, and hence of this world, can be done in a number of ways. We can, for example, postulate that the line consists of points, infinitely small parts, which go to make up the segments of the line. Or, we can think of points as simply an abstraction, a logical construction out of convergent sets of line segments. On the second account, there are no points, except in this derived sense; nothing is infinitely small. A third type of account is another line segment account, but one which says that there are only line segments with rational end points.

Each of these three descriptions of what the world really and at the most basic level consists in can account for all the phenomena and all the geometry of the line. Each account is, in that sense, an equivalent description. But each of these tells different stories about what the straight line world is really composed of. The first, but not the second and the

    [5] In *Meaning and the Moral Sciences* (Routledge & Kegan Paul, London, 1979), pp. 130-3.

third, envisages the existence of points which, being infinitely small, are in themselves invisible and undetectable. The third, but not the second and first, thinks that there are only denumerably many basic objects; the first and the second accounts envisage the line consisting of non-denumerably many components.

Is all this just a matter of the same world being mapped in different ways, Putnam asks, as we can map our world with different projections, such as Mercator or Polar? The answer to this question turns out to be not so simple. We can, if we want, think of these different accounts as different mappings of the same linear world, but if we do, we lose the ability to say anything intelligible about *how* this world is at its basic level. The property of *being an object* in this world, as opposed to being a construction out of classes of objects, becomes relative to the description or account we are using, and so does the cardinality of the world, whether at the basic level it contains denumerably or non-denumerably many objects. What remains constant is the phenomenon, the appearance, but the nature of the unobservable world behind the appearance and its inhabitants is elusive, forever theory-bound.

Putnam's example is, of course, geometrical. There is no suggestion that points or line segments cause the behaviour of actual lines. But, as he says, the underdetermining of our theoretical accounts of how things are by the observable data obtains in similar ways and to the same extent in theoretical physics: 'actual physical theory is rife with similar examples. One can construe space-time points as objects, for example, or as properties. One can construe fields as objects, or do everything with particles acting at a distance (using retarded potentials).'[6] And Larry Laudan gives many other similar examples from the history of science, where conflicting accounts of what lies behind the world of appearance each did equally well in explaining the appearance, and were for a

6 Putnam, op cit., p. 133.

time, in Putnam's terminology, equivalent descriptions. These examples include the debates between Copernican and Ptolemaic astronomy from 1540 to 1600, between Newtonians and Cartesians from 1720 to 1750, between wave and particle optics from 1810 to 1850, and between atomists and anti-atomists from 1815 to 1880.[7]

It may be that the examples given by Laudan were not forever empirically equivalent, and that, in the end, observational tests were devised which enabled decisions to be made in favour of Copernicus, say, and against Ptolemy. Two points, though, need to be made to this response. First, the possibility that, at some future time, currently equivalent theories may be firmed up, and allow empirical comparisons to be made, does not give us any guidance at the present time, nor, secondly, does it offer any guarantee that our future interaction with the world will at some time allow us to discriminate empirically between them. There may be occasions where we could tell even now that this is actually highly improbable, owing to the relationship between the disputed claims and the evidence, including *any* possible evidence. Bas van Fraassen illustrates this point with reference to axiomatic treatments of Newtonian mass.[8] In Newton's own treatment of mass, each body has a definite mass all the time, but there are other axiomatizations of Newtonian mechanics which define mass in such a way that, for unaccelerated bodies, either no mass is defined or they are all arbitrarily assigned the same mass. What we have in such a case is a radical and ineliminable underdetermination of theory by data, because no actual or possible observation can settle the issue. In the Newtonian case, genuine physical differences are undetectable observationally; whether the mass of an unaccelerated body is what Newton says it is, or

[7] Cf. L. Laudan, *Progress and Its Problems* (Routledge & Kegan Paul, London, 1977), pp. 47–8.

[8] Cf. Bas C. van Fraassen, *The Scientific Image* (Clarendon Press, Oxford, 1980), pp. 60–3.

whether it is equal to the mass of all other unaccelerated bodies, surely represents a definite physical difference, but an empirically undiscoverable one. And there are analogous cases in quantum mechanics and relativity theory, as van Fraassen points out.

In the last few paragraphs, we have shifted from talking about unobservable entities and processes to thinking about the underdetermination of theory by data, and considering how, on occasion, the observable evidence does not enable us to choose between competing explanatory accounts of the data. It may initially seem that the two questions are rather different. In the case of unobservable entities it seems that more powerful instrumentation might produce observability where once there was none, whereas the notion of the underdetermination of theory by data implies that, as with the interpretations of Newtonian mass, there can be physical differences which are in principle undetectable. And yet, on examination, unobservability and underdetermination turn out to be not so distinct in practice. Any observational basis is going to be consistent with different stories about its underlying nature. These stories may, on occasion, refer to unobservable entities and forces, and it may be that our observational powers are never improved enough to give us more than the most indirect evidence for these entities and powers. We have already seen Mary Hesse ready to concede that electrons may not exist. Something, or perhaps some things, do what (we think) electrons do. But if we cannot now get closer to the electron than the observation of things done by electrons, the same will go with a vengeance for other more elementary particles. We may never get any closer to these things, and we may have to accept the possibility that, in their case, some other theory may in time be developed which postulates a quite different world underlying the observable world.

When we come to talk of forces, as opposed to talk of entities, however, the underdetermination of theory by data has long been known, and the way in which it was discussed by Berkeley and his successors comes very close to expressing the radical doubts we found in considering Newtonian mass. Berkeley regarded Newtonian talk of absolute space and time, and of force and gravity, as otiose. Absolute space and time are, according to Berkeley, unobservable. We have knowledge of motion and position only as relative to the movement and position of other bodies, and in any case we have no need of an absolute space or time, for we can do all the measuring we need to do by reference to things we can observe, such as the fixed stars. Postulating an absolute space and time will, in any case, help us not at all with actual measurement, as no such thing can be observed.

It will be obvious that Berkeley's position on the relativity of measurement has been vindicated with a vengeance in relativity theory. No less significant is the Einsteinian support for Berkeley's position on gravity. According to Berkeley, Newton's introduction of the force of gravity into his system was tantamount to the introduction of an occult metaphysical substance. All we can actually observe are the effects of this substance, planets orbiting the sun, the tides varying with the moon's position, apples falling to the ground. In introducing the idea of gravity, Newton was able to deduce correct predictions for these, and for many other phenomena. But let no one think that, in doing so, he had revealed their underlying nature or essence. Berkeley writes in his *De Motu* that

'Force', 'gravity', 'attraction' and words such as these are useful for purposes of reasoning and for computations of motions and of moving bodies; but they do not help us to understand the simple nature of motion itself, nor do they serve to designate so many distinct qualities . . . As far as attraction is concerned it is clear that it was not introduced by Newton as a true physical quality but merely as a mathematical hypothesis.[9]

[9] G. Berkeley, *De Motu* (1721), section 17.

In other words, for Berkeley an invisible force like gravity was not to be regarded as anything real. The notion of gravity was part of a theory which allowed mathematically precise predictions of things which could be observed but was not to be regarded as itself referring to anything outside the theory. Of course, this attitude to unobserved causal links paves the way for the Humean analysis of causality, in which all talk of causal relations pertaining between events is nothing more than a disguised way of referring to regularities (or constant conjunctions) existing between types of observed event. Causality itself, on this account, becomes no more than observed regularity, and the difference between what we might think of as a genuine causal regularity and a mere chance regularity is, as A. J. Ayer puts it in his study of Hume, only a 'difference in our respective attitudes towards them'.[10] It is not founded in any physical necessity or necessary connection inherent in the essences or nature of things. In the Humean account, indeed, there are no such things as physical necessities or necessary connections, and any regularities we do observe could, of course, fail to obtain in other circumstances.

Some may be surprised to read Berkeley's claim that, for Newton himself, gravity is not a true physical quality but merely a 'mathematical' hypothesis, one way among many possible ways of calculating and predicting the observed phenomena. After all, was not the existence of gravity one of Newton's great discoveries? In the light of Berkeley's comments, though, Newton's own words in the General Scholium at the end of Book III of the *Mathematical Principles of Natural Philosophy* are of interest:

We have explained the phenomenon of the heavens and of our sea by the power of gravity, but have not yet assigned the cause of this power. . . . It must proceed from a cause that penetrates to the very centres of the sun and planets, without suffering the least diminution of its force; that operates not according to the quantity of the surfaces of the particles on which it acts (as mechanical causes used to do), but according to the quantity of solid matter which they contain, and propagates its virtue on all sides to immense distances, decreasing always as the inverse square of the distances.

After this elegant statement of the inverse square law, Newton goes on to say 'But hitherto I have not been able to discover the cause of those properties of gravity from phenomena, and I frame no hypotheses; for whatever is not deduced from the phenomena is to be called an hypothesis; and hypotheses, whether metaphysical or physical, whether of occult qualities or mechanical, have no place in experimental philosophy.' One naturally feels like saying that gravity is itself the cause of those properties of gravity, but Newton himself is more cautious and more sophisticated, aware no doubt, as Berkeley was, that many empirically equivalent hypotheses about unobservable underlying causes may be consistent with the phenomena, and none strictly deducible from them.

Newton goes on to draw a crucial distinction between what he calls hypotheses (i.e. theories about the unobservable) and what are sometimes called phenomenological laws (theories which do no more than register regularities among observable effects). He then says:

In this (ie experimental) philosophy particular propositions are inferred from the phenomena and afterward rendered general by induction. Thus it was that the impenetrability, the mobility and the impulsive force of bodies, and the laws of motion and gravitation were discovered. And to us it is enough that gravity does really exist, and acts according to the laws which we have explained, and abundantly serves to account for all the motions of the celestial bodies and of our sea.

But does the *existence* of gravity, as interpreted by Newton, amount to more than the observation that certain regularities in accordance with the inverse square law obtain?

Whatever Newton's own final stance, it is clear that gravity need amount to no more than this observation. And if this were all that had been meant, Newton would have been wise, because the same phenomena which he explained by invoking gravity have been explained by Einstein and by more recent theorists in terms of the curvature of space around material bodies, and without invoking a special force at all. Underdetermination clearly applies here because the phenomena are in each case consistent with conflicting underlying explanations.

In this section on unobservability and underdetermination, we have seen that, once we go beyond the observable world in science, problems arise as to the existence and nature of the entities and processes our explanations postulate. These problems arise not simply because we are speaking of things we cannot observe, but more because there will be ever so many possible 'mathematical' hypotheses, all consistent with whatever data we are taking as our observational basis. At this point in science, there will be a critical underdetermination of theory by data, and this in itself seems sufficient reason for holding on to some distinction, however rough and ready, between observation and theory. In the next chapter, we will look more closely at other aspects of the significance of scientific attempts to penetrate beneath the appearances.

6

Scientific Realism

Positivism

At the end of the last chapter, we began to look at the difficult problems which arise when the theories of science postulate the existence of unobservable entities and forces. In this chapter various reactions to the postulation of unobservables in science will be considered, starting with the approach known as positivism. This approach pushes further the scepticism about the unobservable we picked up in Berkeley and Hume, and indeed it may be regarded as the spiritual heir of their empiricism.

Ian Hacking has usefully identified six key ideas characteristically associated with a positivistic approach to natural science.[1] Although not all philosophers who accept one or more of the ideas automatically accept all six, the six taken together do constitute a consistent and coherent position, and one which specifies a particular attitude to science and its problems.

The first idea is an emphasis on verification and/or falsification. The point here is that a significant theory about the physical world should make some difference to experience. It should specify or predict observable states of affairs, and it should be capable of conflicting with observable evidence. This might seem at first sight to be a fairly minimal require-

---

[1] In his *Representing and Intervening* (Cambridge University Press, 1983), pp. 41–57.

ment, until we reflect on the number and extent of influential ideas which fail to meet it, from religious dogmas through political ideology and ethical beliefs to important and fruitful ideas in science itself. Into this last category will fall not only obviously metaphysical propositions, such as 'Every event has a cause', but also ideas which have suggested something about the basic form of the natural world, without being themselves either verifiable or falsifiable. One may think in this context of atomism (the idea that at bottom the world consists of tiny indivisible particles and their billiard-ball-like movements and configurations) or of mechanism (the idea that there is ultimately no action at a distance, and that all physical processes resemble the workings of clocks). Atomism and mechanism share with determinism the properties of unverifiability and unfalsifiability. That is to say, we can never prove that they are true, for fairly obvious reasons, but neither can we falsify them. Even if a leading current theory is indeterministic or anti-corpuscular or unmechanistic, one may dream that, in the end, causes, atoms, and mechanisms will be found. And such dreams have inspired much positive research, and have led to theories which have been empirically verifiable and falsifiable. So the positivist cannot simply say that all non-verifiable, non-falsifiable theories are without significance, even in the context of natural science. But the account the positivist is inclined to give of such theories will tend to stress their heuristic and hortatory significance, and to play down their claims literally to represent the world.

The second positivist idea noted by Hacking is the thesis that sensory observation founds all genuine knowledge. Traditionally, positivism has often been associated with attacks on religion and metaphysics, and attempts to show that, in contrast to natural science, such enquiries cannot issue in real knowledge. So it would be something of a disaster for the old-fashioned anti-religious positivism if it turned out that natural science made significant use of what appears to be non-observational knowledge and claimed such knowledge. We saw in the last chapter that science does appear to give us knowledge of things and processes which cannot be observed, to say nothing of the influential metaphysical theses we find in the history of natural science. So, whatever broader view may be adopted on religion and metaphysics, the positivist will have to deal to his own satisfaction with the claims to knowledge of the unobservable made in modern science.

The third positivistic idea can be traced straight back to Hume, for it is the claim that talk of causation amounts to no more than talk of constant conjunctions between types of event. As Hacking points out, Newton himself, with his rather ambiguous attitude to the force of gravity, helped to smooth the way for a more philosophically inspired scepticism regarding physical necessity; certainly some of Newton's contemporaries, including Leibniz, saw gravity as an illicitly unexplanatory and inexplicable occult power and rejected it, except as an indirect way of referring to observed regularities in the world.

The fourth idea connects with the positivist hostility to causes. It is a suspicion of the role, and even of the possibility, of deep explanations in science. If there is no physical necessity forcing events to happen, and all we have in the world are mere regularities between types of event, then the most we can do by way of explanation is postulate wider ranging regularities. We can see a single event as part of a pattern of regularity, and our first pattern of regularity as part of a wider pattern. But in this wider pattern, all we will have are brute regularities; there will be no sense that the phenomena in the wider pattern have to be connected, or be, as they are. And if there is no necessity at the deepest level, there will be no necessity at any other level. If people hope that a true or deep explanation in some sense explains why things have to happen, then, says the positivist, that hope will not be satisfied in science, at least, when science is

properly understood. And—some positivists will add—if all
we are ever really going to be offered in science are mere
regularities between types of event, what is so good about
regularities between unobservable events? Why do we not
rest content with observable regularities?

The fifth positivist thesis is perhaps the most characteristic. It certainly is the most controversial and the one most
likely to put most people on their guard. It is a thorough-
going hostility to unobservable or theoretical entities. The
hostility to causes and physical necessity may have softened
some people up for this, but it does seem obvious that the
history of science has continually progressed from the postulation of entities to their observation and manipulation.
Hacking mentions molecules, genes, and viruses. The lesson
would appear to be that if science has, precisely through the
postulation of theoretical entities, brought us to be able to
observe what we could not previously observe, we should not
start with scepticism (or worse) about such things. Is the
history of science not very much a history of the unobservable becoming observable, and a consequent blurring of the
observation–theory distinction?

We will not be able to evaluate the positivist position on
theoretical or unobservable entities fully until we have considered the anti-positivist or realist arguments in favour of
accepting their existence. For the positivist would surely be
right to insist that some argument is needed before we accept
that there are unobservable things, as human experience,
broadly conceived, is our only touchstone of reality. Raising
doubts about unobservable existences is not prima facie
unreasonable. It cannot be compared to global or Cartesian
scepticism. And the positivist need not be dogmatic or
inflexible about observation. He could say that science has
extended our powers of observation in many ways, and
accept much of what he sees through telescopes and microscopes, say, as observed. What he will probably do at this
point is to remain somewhat open-minded about where to

draw the line between theory and observation, whilst insisting with van Fraassen that such things as space-time, fields, elementary particles, and alternative possible states of affairs are definitely not observable,[2] and should not be regarded as on a par with observable entities. In so far as accepting the real existence of such things depends on our accepting certain controversial and highly theoretical theories, our reasons for accepting the entities can be no stronger than our reasons for accepting the theories. In fact, as we shall see in the next section, when we have theories which greatly transcend any possible observational basis, the reasons for accepting either theory or postulated entities may not be very strong.

Hacking's final mark of the positivist is that he is opposed to metaphysics. We have touched on this aspect of positivism several times already, and no more needs to be said about it now, beyond making the obvious point that where the anti-positivist will point to the cognitively significant input metaphysical ideas make to the construction of empirical theories, the positivist will interpret the significance of metaphysics in heuristic terms, as forms of explanation simply useful in guiding empirical research. But he still has a problem in explaining why some forms of explanation, such as atomism and mechanism, have proved highly successful in our investigations of the real world. In fact, as we are now beginning to see, much of the dispute between positivists and their realist opponents hinges on the different conclusions they would be inclined to draw from the fact that a certain theory is, at a given time, a good explanation of the observational data.

[2] Cf. *The Scientific Image* (Clarendon Press, Oxford, 1980), p. 202. Many of the subtler points I make in this chapter in favour of positivism are derived from van Fraassen.

The Inference to the Best Explanation

The scientific realist believes that the theories of science give us knowledge about the unobservable. If his realism is to have any bite, he will not simply believe that the theories of science make statements about unobservable things. He will also believe that we sometimes have good reasons for believing that those statements are true.

What reason could we have for believing that statements about unobservable entities and powers are true? The evidence here will obviously be indirect as we cannot directly observe these things. But, says the realist, we may on occasion have good indirect evidence for believing in them, for believing that some of the things we can observe are manifestations of, or effects of, unobservable entities or forces. The majority of arguments of this sort either are versions of what is known as the 'inference to the best explanation', or in some way rest on this.

The thought which underlies the inference to the best explanation is that if a theory explains some data better than any other theory explains them, we have thereby a good reason to think it true. In the inference, explanatory power is taken to be a reason for belief. In Conan Doyle's stories, the explanations of the evidence given by Sherlock Holmes were always more complete and all-encompassing than those initially given by Watson or the police, who usually failed to account for some awkward fact or other. The very completeness of Holmes's explanations *qua* explanations gave them a probative force lacking in the other accounts. And this probative force was characteristically corroborated by a confession from the guilty party or something equally incriminating. The inference to the best explanation was both deployed and corroborated in the stories.

Is the situation in science like that in a fictional crime story? In two highly significant ways, the cases are quite different. In the first place, Conan Doyle is writing fiction.

He is able, by artifice, by hint and emphasis, and, indeed, by the nature of the genre itself, to suggest to the reader that Holmes and only Holmes has taken full cognizance of all the salient evidence. The slower witted doctor and policeman, the reader realizes, always overlook some vital clue, or fail to account for it. We have, in other words, a pretty good sense that Holmes does always provide a better explanation of the mystery, even before his evidence is triumphantly endorsed in the denouement. And, of course, it is endorsed. We get in the end a direct proof or admission of what Holmes had indirectly argued for.

In the scientific case, however, the best explanation is not always so clearly signposted. An explanation which does well in one respect may not do so well in another. An explanation which is simple in respect of the formulae used or the number of basic types of entity postulated may not be very accurate. On the other hand, an accurate explanation may be very complicated and difficult to apply and understand. Also, various explanations may make different estimates of just which salient bits of evidence need explaining.

But even if we could unambiguously and uncontroversially establish that one explanation of a given phenomenon was the best of all available competitors, we would still, in respect of the unobservable entities postulated, lack the direct Holmesian admission or confession. The unobservable entities would only be inferred in virtue of the explanation they are a part of. And how strong an argument is that in their favour? We are left with the feeling we noted in connection with scientifically influential metaphysics, that it cannot be merely coincidental that a particular metaphysical or theoretical account actually fits the facts so well. There *must* be some truth in it, over and above its just fitting the observable facts. Popper writes in this vein of 'realists' who

not only assume that there is a real world but also that this world is by and large more similar to the way modern theories describe it than to the way superseded theories describe it. On this basis, we can argue that it would be a highly improbable coincidence if a theory like Einstein's could correctly predict very precise measurements not predicted by its predecessors unless there is 'some truth' in it.[3]

What those who, like Popper, are inclined to argue in this way need to spell out is just what this 'some truth' in Einstein amounts to. Obviously, what is intended is that there must be some truth at the deep theoretical level: that its ability to make correct predictions of very precise measurements underwrites its picture of the inner and unobservable workings of the world. The positivist, on the other hand, will argue that even very good results at the level of empirical predictability in themselves give us little reason for belief that the world is just as the predicting theory says it is. We cannot, in other words, argue from explanatory power to truth.

The positivist position can seem impregnable once it is realized that conflicting explanations at the deep theoretical level can be equivalent at the level of empirical observability, that they can be empirically equivalent, in other words. And for this to apply, we do not need actual examples of conflicting explanations. Their mere possibility is enough to show that one cannot go directly from explanatory power to truth, because we could get another and inconsistent theory to explain the same data on a completely different theoretical basis. And this will be true whatever we take to be our empirical basis, so the point is not affected by problems involved in the precise demarcation between the theoretical and the observational.

While, from an abstract point of view, the positivist position can seem unassailable, a natural reaction here is that in science we are not involved in mere abstractions. We are concerned with specific highly complex and concrete theories, of a type which have had and continue to have great

[3] K. R. Popper, 'Replies to Critics', in P. A. Schilpp (ed.), *The Philosophy of Karl Popper* (Open Court, La Salle, Ill., 1974), ii. 1192–3.

success in dealing with the empirical world, and in leading to all sorts of discoveries in that world. These theories are in many ways continuous with our considerable empirical knowledge of the world and seem to flow seamlessly from it. The very problem just alluded to, of demarcating the observational from the theoretical, suggests that there is something highly artificial in, for example, regarding what we see in a powerful light microscope as observed, while disallowing evidence from electron microscopes as 'theoretical'. And what are we to say of evidence gleaned from polarizing microscopes, in which certain coincidences between light rays and particular cell fibres are exploited to allow us to 'see' certain normally transparent fibres of living organisms?[4] The point here would be that it is legitimate to use any relevant property of any kind of wave, including even acoustic waves (as we see in the case of ultrasound), in order to create images of structural features of organisms and other entities. If real properties are imaged in this sort of way, who is to say that observers are not *seeing* what is presented?

The fluidity of the distinction between what is really observed and what is merely inferred is, and is likely to remain, the Achilles' heel of the positivist. In his defence, two points can be made. He is not committed to maintaining an absolute and timeless distinction here, nor to saying that the unaided evidence of our senses is our only way of observing the world. He can certainly allow extensions and corrections of our unaided senses by means of instruments, nor are such extensions to be confined to cases like those of Jupiter's moons, in which Galileo's telescope might be thought of as simply anticipating things which were in principle humanly observable without artificial observational aids. Molecules and cells are never going to fall into that category, and even less are chromosomes or electrons; so, if

[4] On this point, cf. Hacking's *Representing and Intervening*, at p. 197. On microscopes generally, cf. the whole of the relevant chapter in this work.

we can observe them at all, it is only going to be with instruments.

Does this mean that, according to the positivist, such things are forever unobservable, and hence, forever only inferred or theoretical? The positivist, surely, does not need to say anything so extreme. He can say, as Hacking does in his chapter on microscopes, that the observable realm can grow through instrumentation, when certain conditions are satisfied. Basically, what we want when observing through instruments, as with any other form of observation, is an assurance that what we observe is caused by reality, and not due to aberrations of our instruments. And we do get some assurance of this sort when we find that physically completely different instrumental techniques all come up with much the same picture of the reality observed. As Hacking says, 'light microscopes, trivially, all use light, but interference, polarizing, phase contrast, direct transmission, fluorescence and so forth exploit essentially unrelated phenomenological aspects of light'.[5] The conclusion is that it would be a 'preposterous coincidence' if two (or more) completely different kinds of physical systems were to produce the same arrangements of phenomena on their screens or micrographs or whatever.

Now this argument from cross-checking is not the inference to the best explanation. As yet no explanation or account of the microscopic phenomenon which causes the variously observed effects is at issue. What we have here is simply an extension of the way we appeal to cross-checking of sensory sources in the macroscopic world. Our sense of sight is corroborated, and at times corrected, by our sense of touch, and the argument gets stronger when we are able to interfere in predictable ways with the things observed. Evidence from different senses and different observers is mutually reinforcing and correcting, in such a way as to lead

[5] Ibid., pp. 203–4.

to a good sense of those occasions on which what we observe is due to aberration on the part of our sensory organs, rather than produced by direct interaction between an object and, as Hacking puts it, 'a series of physical events that end up in an image of an object'.[6]

Given relevant cross-checking, then, there seems little reason in principle to refuse to say that one might on occasion even see by means of an acoustic microscope, in which bursts of sound are projected on to an object, and the responses transformed into an image on a screen. The seeing here may, in a sense, be indirect, but no more indirect than that involved in watching television, or in a mother seeing her foetus on a screen in front of her, by means of ultrasound. In all these cases, the image tracks the object. The image is caused by and varies according to the behaviour of the object. If the object were different or behaved differently or if we interfered with it in some way, the image would vary in relevant ways. If we are prepared to allow ultrasound observation in the case of a foetus, there seems little reason in principle to disallow the possibility of acoustic microscopes, providing their deliverances can be cross-checked by different types of light microscopes.

It is true that one is able to 'see' better when one has some idea of what it is one is seeing, and some idea of how what it is that one is seeing functions. Hacking points out that the early electron micrographs of genes could not properly be recognized as such before there was some conception of what functions were played by the various bands and interbands on the chromosome.[7] And, as he adds, the same would be the case for a Laplander stranded in a Congolese jungle, who would make very little sense of what he saw around him. If we say that our assurance that we are really observing something increases when we have an explanation for what it is we are observing, we are not thereby committed to the

---

[6] Hacking, *Representing and Intervening*, p. 207.   [7] Ibid., p. 205.

inference to the best explanation, nor to thinking that our explanation receives any great support from what it is we now believe we are observing. For, as we noted earlier in connection with Kuhn, phenomena, once established, can survive changes of explanation, even changes of those very explanations which were instrumental in originally picking out and identifying the phenomena. Explanations and theories may well be helpful in leading us to organize our data, and observe our world. But explanations and theories which are not true can play this role, and once identified via some theoretical perspective, a given phenomenon can become well enough established through cross-checking and manipulation to survive abandonment of the original explanatory framework.

What I am trying to suggest here is that the positivist—or, at least, someone who finds much of the positivism outlined in the previous section persuasive—can go quite a long way in allowing that we can observe many things which we cannot perceive with our unaided senses without giving up his suspicion of full-blooded scientific realism. In particular, he does not have to accept the inference to best explanation, because we do not necessarily need this inference to establish the existence of sensorily unobservable phenomena. What the inference to best explanation attempts, controversially, to do is to establish the truth of explanatory theories just where they go beyond the evidence, however the evidence is conceived. And this leads us to our second consideration in favour of the positivist. The fact that the boundary between the observational and the theoretical is not clear cut, and that it may shift with improved instrumentation, does not show that the positivist is wrong to think that there are some cases which definitely fall on the theoretical side of the line. What we want to know is what we are to say about those entities and forces mentioned in scientific theory which cannot be observed at all, but are at best inferred from their effects (according to our theories). And this question remains pressing now, even if at some later stage we might get more direct observational knowledge of them.

Van Fraassen, in his attack on it, takes the inference to best explanation to be of the form: where we have evidence E, and hypotheses H and H', we should infer H rather than H' if and only if H is a better explanation of E than H' is.[8] Van Fraassen mounts a powerful case against any such inference, and hence against belief in any entities established *only* by the inference. But his case, which we will now examine, would not count against (initially) unobservable entities if (as I have just suggested) there might be other ways of establishing their existence. Van Fraassen himself is undoubtedly far less tolerant than we have been of any of the type of extension of observation we have been suggesting, but in a way this is beside the point. His attack on the inference to best explanation still raises considerable problems for anyone tempted to full-blooded scientific realism.

What van Fraassen argues is that all we are entitled to assert by any evidence which supports a hypothesis about the unobservable is that things are *as if* there were forces or photons or electrons, or whatever the unobservable entities in question are. The mere fact that postulating unobservable entities provides some explanatory account of some observed regularity is in itself no reason for belief in the entities. Even if our explanatory hypothesis is the best available, the evidence actually entitles us to say no more than things behave as if they were brought about by whatever our explanatory model postulates.

At this point the realist will probably introduce two further considerations. He will bring in the improbable coincidence argument and also assert that any natural regularity needs explanation, if necessary at the unobservable level. Van Fraassen is able to make rather short work of the second point, showing that such a demand for explanation leads to absurdity. If we 'explain' our observed regularities

[8] Van Fraassen's positivistic attack on the inference to the best explanation is in *The Scientific Image*, pp. 19–40.

by postulating unobserved forces and entities, these too will act according to the regularities mentioned in our deep theories, and so themselves need some further explanation. The feeling van Fraassen evinces is that if one is going to be stuck with brute regularities, we might as well stick at the level of observed regularities, rather than pushing our demand for explanation unnecessarily down into the unobserved level.

Van Fraassen is also able to show that an unrestricted demand for explanation of regularity, when it takes the form of the search for a common cause of a particular regularity, falls foul of orthodox quantum physics. The types of instantaneous and correlated action at a distance discussed in the Einstein–Podolski–Rosen paper of 1935 do in fact exist, and are not, in terms of orthodox quantum physics at least, further explicable. That is to say, in quantum physics we get cases where two particles acting randomly *and* independently of each other actually mirror each other's behaviour. One such case has been demonstrated by A. Aspect in a series of experiments reported in 1982.[9] Pairs of photons are emitted from opposite sides of calcium atoms. Each travels via a rapidly changing switch to one of a pair of polarization filters beyond. The switches are not correlated, so only random coincidences in the ultimate behaviour of pairs of particles would be expected. But if one particle reaches a given filter on its side, the chances that the other reaches the corresponding filter on the other side are far greater than they should be on the assumption that the behaviour of one particle is independent of that of the other. While we would naturally want to say that there must be a common cause to explain such a regularity, orthodox quantum theory does not permit the postulation of some hidden variable to explain this puzzling phenomenon nor can we think of the particles communicating with each other faster than the speed of light.

[9] A. Aspect, J. Dalibard, G. Roger, 'Experimental Test of Bell's Inequalities Using Time-varying Analyzers', *Physical Review Letters*, 49 (25), 1982, pp. 1804–7.

In van Fraassen's view, there may be a point in working with explanations of observed regularities, interpreting the regularities as if they were brought about by some unobservable cause. Such explanations can lead us to discoveries of further regularities and entities at the observable level—and, we could add, to pushing back the acceptable limits of observation. But if an explanatory theory does this, would it not then be a highly improbable coincidence if it were actually not true? In other words, does not a theory which, through its explanatory account of reality, actually leads us to more empirical knowledge, gain some degree of confirmation or truth-likeness, by virtue of that very fact? The realist is arguing here that the success of the theory is to be explained in terms of its truth or probability, meaning that what it says about the unobservable world has a good chance of being true.

The positivist will, of course, reply that the most that can be derived from knowledge that a theory is empirically adequate and knowledge-increasing is that it has these properties, and that the world is such as to be as if it were as the theory states. Without assuming some deep isomorphism between the mind and the world, we are entitled to conclude no more. And we can explain the success of theories in science without supposing any such thing. Van Fraassen writes about this in Darwinian vein: 'the success of current scientific theories is no miracle. It is not even surprising to the scientific [Darwinist] mind. For any scientific theory is born into a life of fiercest competition, a jungle red in tooth and claw. Only the successful theories survive—the ones which *in fact* latched on to actual regularities in nature.'[10] This Darwinian metaphor is, in fact, worth taking rather further than van Fraassen does.

A scientific theory can be compared to a species produced by biological evolution in that both can be regarded as

[10] *The Scientific Image*, p. 40.

embodying attempts to cope with and anticipate the environment, to fit with it at the empirical level, so to speak.
Successful theories, like members of successful species, do
this of course, or at least they both do it enough to survive.
But they both do it without any direct instruction from the
environment. That is to say, neither the genetic structure nor
the inner core of a theory are directly moulded by the
environment, and this is true even though scientific theories,
unlike biological mutations, are guided by rational considerations. Even though the rational and problem-solving aspect
of scientific theories tends to be overlooked by those pressing
analogies between science and biological evolution, scientific
theories do have a strong element of guesswork about them.
They may not be completely blind and undirected, like
biological mutation, but both can be regarded at least to
some extent as leaps in the dark. Both are guesses, which will
be weeded out if they are too wide of the mark empirically.
But, precisely because they are leaps into unknown—random
genetic sports, initially, or imaginative leaps, in the case of
theories—their fitting the environment reasonably well at the
points of empirical control presupposes no mirroring of the
environment at the level of the genetic code or at that of the
theoretical model of the environment. Equally, lack of isomorphism between the theoretical core of a theory and the
underlying structure of the world will not preclude accurate
anticipations, in either species or theory, of initially unforeseen regularities. In fact, as van Fraassen hints, competition
between theories or species may help fairly quickly to pin-
point those theories or species which are able to anticipate
new actual regularities. The 'some truth' Popper and other
realists are looking for in their successful theories can, not
implausibly, be seen as this ability to anticipate as yet
unknown facts and regularities. But, once again, this ability
can be detached from the truth of the theory at the theoretical level, or precise mirroring of the world by the theory at
that level.

This conclusion could reasonably be resisted by the realist if there were greater theoretical convergence in science than in fact there is. Thinking of scientific theories as if they were biological species might seem to some to be misleading in that we think of science as progressing, while this is not the case in biological evolution. Moreover, biological species often occupy different ecological niches and so are not in direct competition, whereas scientific theories about a given area will be in direct competition. But if we confine our comparisons to scientific theories and biological species in direct lines of competition, there are close parallels. Assuming (which perhaps we shouldn't in the biological case, except for the point of our example) that an ecological niche has not altered too much over time, we might want to envisage later occupants of the niche as improving on the fit of their predecessors, as succeeding where their predecessors have succeeded in fitting the environment and in some other places as well. But to envisage this type of better fit in later occupants of a given niche, we do not have to envisage any very close resemblance between them and the earlier occupants they may have displaced. Indeed, they do not even have to be of the same line of descent.

Something like this piece of armchair biology actually represents the situation in science, and provides the biggest stumbling-block for those who would espouse full-blooded scientific realism. Explanations which were once successful have come and gone, but in their going, their explanatory core—their pictures of what the world is fundamentally like—have often gone too. The empirical knowledge they produced has remained, but without the explanatory accounts. No doubt the inference to the best explanation could have been applied in the past, with great rhetorical effect, to Newton's theory, say, or to atomism. But without the perspective provided by the idea of convergence of knowledge through scientific revolutions, its effect is rhetorical, rather than genuinely persuasive. What price the inference to

the best explanation in our case, when, if the history of
science teaches us anything, it is that even the best explanations of their time have been eventually overthrown, and
many of their theoretical assumptions abandoned with them?

Scientific Laws and the Representation of Reality

In the previous sections we have seen that there are grounds
for distinguishing between the empirical predictions and
discoveries made by scientific theories and their theoretical
core—the model of the universe the theories encapsulate.
Empirical adequacy is no reason for thinking that the model
advanced by a theory is true. We have, though, so far
assumed that we are entitled to think of theories as empirically adequate, as giving fairly precise knowledge at the level
of empirical measurement. In this section, we shall see that
there are reasons for taking even this to be a somewhat
idealized view of scientific theories, which, it turns out, even
at the empirical level are often idealized abstractions.

In one very obvious sense of the term, it is clear that many
scientific theories are idealizations. Take, for example, Newton's laws:

(1) every body continues in its state of rest or uniform
motion in a right (i.e. straight) line unless it is compelled to
change that state by forces impressed upon it;

(2) the change of motion of a body is proportional to the
motive force impressed, and is made in the direction of the
right line in which that force is impressed;

(3) to every action there is always opposed an equal
reaction, or, the mutual actions of two bodies upon each
other are always equal, and directed to contrary parts.

But Newton's law of gravity states that the attractive force
between two bodies is proportional to the product of their
masses divided by the square of the distance between them.

From this it follows that no object anywhere in the universe can be free of gravitational forces impressed upon it by other bodies. So no object anywhere in the universe can ever be in the state specified by the First Law.

Is the First Law, then, literally true? If it is, it is certainly a truth we can never check directly by observations. Its literal truth can never be confirmed by observation of even a single case, for no case actually falls under the condition specified, of being under the influence of *no* impressed force. And so we cannot check whether such a body actually does continue in its state of rest or of uniform motion, any more than, as we have already seen, we can check whether unaccelerated bodies have any mass or not. Thus neither the literal truth of the First Law nor the precise interpretation of Newtonian mass can be established by comparing the law or the concept with reality. But, if reality does not determine such things, what are we to say of such laws and concepts themselves? Should we regard them as in the business of giving literally adequate (or inadequate) descriptions of reality? Or should we look at them rather as tools or idealizations or abstractions, more or less useful, within certain limits, for calculating and predicting natural phenomena?

As we noted in Chapter 2, in discussing the achievement of Kepler, modern science has been much concerned with giving mathematical accounts of natural phenomena. Much of its strength, of course, derives from this drive to quantify, and from the successes achieved in the quantification of empirical phenomena. But, how far can we regard the workings of natural phenomena as being really capturable by mathematical formulae? Kepler's own work here was underpinned by his Pythagorean faith in number being the essence of the world. But does experimental work really bear out this faith, or do we rather find that things in reality only approximate to mathematical models, and do not follow them absolutely?

In abstract philosophical discussions of science, the extent to which the laws and theories of physics are frequently idealizations, approximations, and simplifications is often overlooked. What is actually found in nature is far richer and more untidy than our theories assume, but we often ignore or regard as irrelevant those aspects of actual states of affairs which do not match our theories. Mismatches of detail are characteristically attributed to factors extraneous to what we are attempting to cover with precise theories, and which we have been unable to control. We have just seen that there are respects in which Newton's theories do not precisely map the observable world, but it is interesting to note, as Popper points out, that Newton regarded the solar system itself as imperfect when compared with what pure mathematics might have required.[11]

Popper goes on to quote C. S. Peirce, an experimentalist as well as a philosopher, to the effect that even the 'most refined comparisons' of masses and lengths, which far surpass in accuracy the precision of other physical measurements, 'fall behind the accuracy of bank accounts'. The determination of physical constants generally are, in Peirce's view 'about on a par with an upholsterer's measurements of carpets and curtains'.[12] There is, in other words, a tension between the untidiness of reality, and the simplicity and generality we look for in our theoretical accounts of that reality. Accounts which were not reasonably simple and general would not be widely applicable, but the price of simplicity and generality is idealization and approximation even at the observable level. Our theoretical accounts of nature often apply perfectly only in ideal and controlled situations. To apply them to real environments, we modify them in various ways, through so-called bridge principles, which tell us how to apply a theory in specific circumstances, through *ad hoc* corrections and the like. In order to manipulate nature and to

[11] Popper in *Objective Knowledge* (Oxford University Press, 1972), p. 212.
[12] Cf. Peirce, 'The Doctrine of Necessity', in his *Collected Papers*, Vol. 6 (Belnap Press, Cambridge, Mass., 1935), pp. 35–65, at p. 35.

get some unified view of how things work in a particular
natural domain, we envisage a nature consisting of clearly
demarcated natural kinds, all the members of which obey the
relevant laws by virtue of their common essence. The virtue
of having such an account is obvious, as we suggested at the
end of Chapter 3, particularly if our account more or less
approximates to how things are. But we can have no a priori
assurance that nature is really like that, and that it is not
altogether more fuzzy and irregular.

The claim that nature is to a degree fuzzy and irregular has
recently been defended by Nancy Cartwright in a collection
of essays appropriately entitled *How the Laws of Physics
Lie*.[13] She gives examples of how, as she puts it, facts are
fitted to equations. In quantum mechanics, for example, free
particles are represented as plane waves. Such a wave would
naturally go all the way to infinity. But the square of the wave
at a given point is what represents the probability of the
particle being at a given point, so the integral of the square
over all space must equal one. This, though, is impossible if
the wave goes to infinity. One widely accepted way round
this difficulty is to assume—contrary to fact—that the walls
of the tube in which the particle is enclosed produce an
infinite potential, from which it will follow that the particle is
no longer to be seen in terms of an infinite plane wave.

Cartwright comments that we know that it is the case that
there is a probability of one of the particles being in some
finite region of space, while it is not literally true that the
walls interacting with a particle produce an infinite potential.
But 'it is not exactly false either. It is just the way to achieve
the results in the model that the walls and environment are
supposed to achieve in reality. The infinite potential is a good
piece of staging.'[14] Earlier she had spoken of the way Thu-
cidydes conceived his task in writing history, as not being to
perform the impossible feat of reproducing exactly what was

[13] Clarendon Press, Oxford, 1983.   [14] Ibid., p. 142.

said on a given occasion, but rather to represent historical agents as saying things which would convey the spirit of their sentiments. She likens this to a dramatic representation of a historical event, in which the conventions of the stage impose limits on the literalness of one's depiction of character and action. Characters whispering to each other in reality have on the stage to speak up, so that the audience might hear what they are saying. This sort of thing is not exactly true to life, but it is not exactly false either, being the natural way to represent an event in a given medium. And it is for a similar sort of reason that, in scientific representation, we may fit facts to our equations or models.

More generally than this conscious fitting of facts to equations, there are also, according to Cartwright, many occasions in practice where actual objects or situations do not obey the laws they are supposed to obey. She takes the example of Maxwell's radiometer, a little windmill, whose vanes are black on one side and white on the other, which is enclosed in an evacuated glass bowl. The vanes rotate when light falls on the radiometer, and the rotation is ascribed to the action of the gas molecules left inside the bowl. Rather in the spirit of Peirce, Cartwright looks at Maxwell's mathematical treatment of the distribution of the gas molecules which is supposed to apply where in the radiometer there are inequalities of temperature and viscosity, and where the viscosity varies 'as the first power of the absolute temperature', as Maxwell puts it. But, says Cartwright, these conditions are not met 'in any of the radiometers we find in the toy department of Woolworth's. The radiometers on the shelves of Woolworth's do not have delicate well-tuned features. They cost $2.29. They have a host of causally relevant characteristics besides the two critical ones Maxwell mentions, and they differ in these characteristics from one to another.'[15] In contrast to the smooth picture of science we are

[15] Ibid., p. 154.

often presented with, in which there are subsidiary laws or bridge principles taking us down in a law-like way from idealized general theories to specific cases, Cartwright says that we know of no principles which will determine how or why Woolworth's radiometers deviate from Maxwell's function for the distribution of their gas molecules. Nor is it the case that even those radiometers which do meet Maxwell's conditions obey his function. Most of them have many other relevant features affecting their behaviour. Cartwright's conclusion is that, in the absence of any laws linking the ideal case to real cases, we have to regard Maxwell's distribution function as a pure fiction, with no explanatory force. It is a mere property of convenience, which we have no idea how to apply outside 'the controlled conditions of the laboratory, where real life mimics explanatory models'.

Cartwright's own view is that when in science we want to apply mathematical theories to reality, we have to use fictional theoretical descriptions (or 'models') of phenomena. These models never exactly fit reality. Their equations do fit the objects postulated in the model, but this is because a model is made to conform to the equations. Different and incompatible models may be used for different purposes. In this context, Cartwright cites a text on quantum optics which mentions various different and mutually incompatible models available for lasers: idealized interacting models, soluble models, simplified dynamical models. In each model, the same phenomenon is explained in a different way, and we select our model according to the properties of the laser we are interested in.

Cartwright draws the conclusion that

in general, nature does not prepare situations to fit the kinds of mathematical theories we hanker for. We construct both the theories and the objects to which they apply, then match them piecemeal onto real situations, deriving—sometimes with great precision—a bit of what happens, but generally not getting all the facts straight at once. The fundamental laws do not govern reality.

What they govern has only the appearance of reality and the appearance is far tidier and more readily regimented than reality itself.[16]

If there is any truth in what Cartwright says—and she certainly provides a good deal of supporting evidence—it seems that we should regard the theories science actually provides us with as far from complete and precisely accurate representations of reality. They are idealizations and abstractions which focus on particular properties of natural phenomena and cases of partial regularity, corresponding no doubt to specific interests and concerns we might have. But in applying our models, we overlook both their incompleteness and their inaccuracy. They do well enough for what we want in predicting and controlling effects, but this 'enough' could be quite consistent with a good deal of inaccuracy and a good deal of overlooking of the full detail of any actual situation. Moreover, we choose our models according to the specific features of the situation we may be interested in, without worrying too much about whether one model can easily be combined with another model we might use for other purposes. None of this militates against the idea that science can discover genuine regularities or new phenomena or new entities. But it does militate against the thought that in science our aim is always the production of ever more general and comprehensive accounts of the whole of a given level of existence, which at the same time are ever more accurate. This ideal may be unattainable. It certainly will be if nature is basically untidy and cannot be divided into clearly demarcated natural kinds. And it may be that most of what we want from science, in the way of the explanation and of the control of nature, can be achieved without assuming the validity of the ideal.

---

[16] Cartwright, *How the Laws of Physics Lie*, p. 162.

The Absolute View of the World

Underlying the various disputes we have been considering between positivists and their realist opponents is a fundamental divergence of opinion about what science can do. The dispute is not about the existence of a real world apart from our knowledge of it. Whatever might have been the position of Berkeley or Hume, this is not an issue as between van Fraassen and Hacking or between Cartwright and Popper, or even, I believe, between Kuhn or Feyerabend and Popper. All these philosophers of science accept as a premiss of their enquiries—as indeed I did in the opening passages of this book—that there is a world independent of we who observe it and live in it, a world which impinges on us in various ways, and a world which scientific enquiry can teach us more about than we could discover by random methods or by sticking to sensory observation unaided by scientific theories or instruments. What I have been calling the positivist or, perhaps better, empiricist tendency in the philosophy of science is not, to repeat what I said in considering unobservable entities, a form of Cartesian scepticism. Cartesian scepticism exploits hyperbolic doubt; it doubts wherever there is not a conclusive proof to guard against all and any possible scepticism. In Descartes's *Meditations*, an evil genius could be confusing me about $2 + 2 = 4$, or I could be dreaming that I am writing this. The contemporary philosopher of science, by contrast, is likely to accept our sensorily and culturally given account of the world and its inhabitants as a starting-point. His questioning is about the extent to which we can go beyond this starting-point in order to gain knowledge of the world as it is apart from our particular human perspectives on it.

The hope that we can transcend our human perspectives on the world is, of course, entertained in modern science right from its beginning. The very idea of the Copernican revolution is to displace what must have seemed unreflect-

ively obvious to all men. Despite Aristarchus of Samos, who
did postulate a heliocentric universe in Hellenistic times, it
must have been very hard for Copernicus's contemporaries
to accept that the earth was not the centre of the universe.
Copernicus asserted that our certainty was derived only from
our particular viewpoint on the earth, and that a wider view
of things would show this. Observers from other vantage
points might locate the centre of the universe elsewhere, but
they certainly would not locate it in the earth.

The Copernican example indicates one type of correction
of human perspective made by science. It shows us that
something we very naturally perceive as being the case is not
the case at all, but it serves to cast no doubt on our perceptual
faculties or their deliverances as such. Presumably from very
many points in space we would be able to see that it was far
more plausible to regard the earth as going round the sun,
rather than vice versa. As with our seeing Jupiter's moons
through a telescope, we can easily entertain the thought that,
from a suitable vantage point, we could observe things as
they really are with our normal sensory equipment, and our
not being able to observe them as they really are is due
merely to accidental facts about our current position in
space. There can, then, be corrections of human perspective
in science which involve no very radical claims about limitations or deficiencies inherent in our sensory apparatus.

On the other hand, as we have already seen, there can be
cases where science takes us far beyond our natural sensory
apparatus. Much of modern science, of course, does this; and
much of our discussion in the last two chapters has centred
around the problems involved with this. We have seen that
there can be good reasons for thinking that we can extend our
powers of observation by means of instrumentation. When
this occurs, it would be perfectly in order to think of science
as providing us with a wider, more inclusive view of the
world than that we are born and brought up with. In so far as
we can transcend our biology and natural language in this

sort of way, we can regard science as providing us with a less relative, more absolute account of the world. It is an account less relative to our own perspectives and viewpoints. It is more absolute in the sense that in science we are able to arrive at entities and properties of entities which are causally more fundamental than the properties manifested by the world as it appears to us.

In fact, in the theories of physiology and perception which have been dominant in natural science since the seventeenth century, it is claimed that the world does not appear to us as it really is. A distinction is drawn between the manifest image (the world as it appears to us) and the scientific image (the world as it is in itself). According to the former, the world consists of objects which are coloured, noisy, hard, or soft to the touch and having a given smell and taste; in science, though, we learn that all these sensory properties of things are not in the things themselves, but are due to the interaction of particles, which lack these properties, with our sense-organs. Thus, colour vision is due to the interaction of colourless photons with our visual cones. The world as it really is does not contain colour as a fundamental property. Colour and the perception of colour are due to the interaction of causally more basic properties of the world, and the same goes for other sensory qualities we perceive.

Looked at from the point of view of the scientific image we can see ourselves as part of a more inclusive world and our perspective on it as just that. Creatures with other sensory faculties and powers might well have a different type of manifest image. In a sense we can easily accept that this is so: it is common knowledge that dogs and bats, for example, can perceive sounds and vibrations we are insensitive to, while being insensitive to much of what stimulates us. Their picture of the world would presumably be quite different from ours. Nevertheless, if we could imagine dogs and bats and other types of creature doing science, we might think of them pushing beyond their manifest image of the world to a scientific image not unlike the sensorily purified scientific image we attain in thinking of colours and the rest as due as much to our make-up as to the way the world is in itself. We thus arrive at the idea of an absolute conception of the world, a conception, that is, of the world as it is in itself, independent of any particular mode of perceiving it, and to which all specific modes of perceptions could be related.[17] Such a conception would in principle be available to perceivers of whatever physical and sensory constitution; it would lay out the causally fundamental properties of the world, and in so doing show the types of perception of specific classes of perceiver as due to those more fundamental processes. We might, too, think of the absolute conception of the world as showing how all the different types and levels of property we observe in the world are all reducible to one basic level of property, to one basic essence of matter and of the natural world. In this way, the absolute conception of the world, though not strictly implying either, would connect fairly naturally with anti-Humean essentialism, and with reductionism in the sciences. The biological may thus be reduced to the chemical, and the chemical in turn to the physical, with the natural kinds in each higher level being analysable in terms of lower level natural kinds, whose nature at the physical level necessitates all else. And so the old philosophical dream, that everything in the world, and all its myriad properties and appearances, are produced simply by the regular rule-governed movements of atoms in the void, would be fulfilled.

No doubt this absolute conception of the world is inspired in part by the account of colour and other sensory properties given by Newtonian and post-Newtonian science, as well as by the Newtonian conceptions of absolute space and time. But the idea of the world, as both in principle open to perceivers of whatever constitution and also prescinding

---

[17] Cf. Bernard Williams, *Descartes: The Project of Pure Enquiry* (Penguin Books, Harmondsworth, 1978), p. 211.

from all conditions and perspectives of perception in order to give us the world as it is, independent of all thought, is a *philosophical* ideal rather than an actual scientific achievement. In certain respects the image of the world presented by modern science prescinds from our particular modalities of perception, as we have seen, explaining their nature and genesis. But the actual scientific image we get at any time is not uncontaminated by all human interests and perspectives. Nor is it clear that there could ever be one unified account of the world which reduces everything to one fundamental level of explanation.

We shall see, later on, the issues involved in thinking that scientific explanation can be reduced to one fundamental level. What has already become clear is that it is hard to see our science prescinding altogether from our human point of view. Even though through instrumentation we can significantly extend our knowledge of the natural world, our theories and postulations must in the end have some bearing on things we can observe, either through instruments or by our unaided senses. And our observation of instruments itself will normally be through our unaided senses. Of course, this does not mean that we are just left with our unaided senses, or can learn only through them. As we have seen, we can go a long way beyond that in the direction of a less humanly relative picture of the world.

The problem arises not so much in formulating less relative accounts of the world, as in checking them when they go beyond what we can observe. It becomes particularly acute when we are dealing with what might be called metaphysical theories as to the basic essence of the world, for, as we have already seen, these theories are not testable, nor do we find convergence at this level of speculation in the history of science. A positivistic, anti-realist attitude to such theories, suspending judgement on their truth, may well be appropriate. If it is, then this is an implicit admission that there are limits to what we can know about the world, given initially by our human constitution and physical location, and, to that extent, it will raise a doubt as to the accessibility of an absolute conception of the world.

Whatever might be said about our ability to grasp an absolute conception of the world, or to aim at such a thing, it is important to emphasize that the practice of science does not require us to think that we are aiming at or attaining this. It may be that the regularities uncovered by our scientific theories are not all-pervasive, and apply only in our region of space and time. It may be, too, that our theories do not disclose genuine natural kinds, with essences necessitating their behaviour. There may, indeed, be no natural necessity in the sense denied by Hume. It may also be that our empirical theories are—as Cartwright suggests—models and approximations conditioned by our interests and specific needs. It may further be the case that there are no ultimate reductions of one level of science to another. But none of this would invalidate the results we have gained or the quest for further theories and regularities. The discoveries we have made and will make could still be seen as subserving the Baconian ideal of relieving man's estate. As for the scientific aim of gaining theoretical knowledge of the world, this will to some extent be achieved through our increasing knowledge of regularities at whatever level we can observe the world, by our discoveries of new entities and phenomena, and so on. Someone sceptical of realism at the metaphysical level of science, of the doctrines of essence and of natural necessity, as well as of our ability to achieve a truly absolute conception of the world, could still be seen as following the Baconian edict that nature must be obeyed before being commanded. He will, though, be open to the possibility that our theories and knowledge of the world may have limits given by our position and perceptual status in the world, that they may not take us to the deep essence of the material world, if indeed there is such a thing, and also that such knowledge as we have may be disjointed and not readily combinable into a single picture. Things we know at one level or in one area cannot directly contradict each other, but that is not to say that their connections and relationships are necessarily unproblematic.

Partial Pictures: Schrödinger's Cat

Partly as an illustration of the difficulty of combining our various theories into a single picture, and partly in order to pave the way for our consideration of probability statements in science, we can consider briefly one notorious difficulty in quantum mechanics. Indeed, in some ways, it is *the* difficulty quantum mechanics poses for us, the problem of combining the quantum-mechanical picture of reality, which speaks of collections of micro-states coexisting at any one time and place, with ordinary or classical descriptions of the physical world, in which we think of things and states of affairs as being at any one time in some definite state, rather than in collections or superpositions of states. Apart from the problem of understanding how the spin of an electron can at some time be both 'up' and 'down', without being definitely either, we have to explain how it is that some macroscopic entity connected in some way to the micro-particle or system is enabled to behave as if the particle is definitely in one or other of the states. This is the predicament posed by Schrödinger's famous example of the cat.

We are asked to envisage this unhappy creature trapped in a closed room which contains a Geiger counter and a hammer raised above a flask of prussic acid. The counter contains a trace of radioactive material which, after one hour, has a 50 per cent chance that one of its nuclei will decay. If this happens, the counter will click, the hammer will fall, the flask will be broken, and the cat will die instantaneously. Unless we look into the room, of course, we will not know at the end of the hour whether the cat is alive or dead. Now, the quantum-mechanical description of the radioactive material does not permit us to speak definitely of the decay or non-decay of the nucleus after one hour. The system is not represented as having a unique value in this respect, but as consisting rather of a superposition of possible values, here, presumably, decay and non-decay.

On the other hand, and perfectly obviously, for an observer who looks into the room after one hour there will not be a superposition of a dead cat and a living cat, and the same will be true for the cat itself, which will be either alive or dead, and not in some half-way state combining both life and death. The difficulty is to see how some essentially fuzzy quantum-mechanical system (a nucleus in a superposition of decayed and non-decayed states) relates to the definite macroscopic state of affairs (cat definitely alive or definitely dead). The point of Schrödinger's example is that in it the definite state of the cat is seen as causally dependent on the indefinite state of a particle, which is apparently consistent with the cat being in either state. We can look at the problem in terms of the chain of events leading from the indeterminate quantum-mechanical state to the definite macroscopic event. How and at what point does the chain firm itself up into something definite? And this is not just a problem of temporal chains, whereby something that was fuzzy eventually becomes definite. Present definite macroscopic states of affairs are now sustained by fuzzy, indefinite quantum states.

As would no doubt be anticipated, there have been a number of attempts to solve the problem of the so-called 'collapse of the wave-packet', whereby a superposition of quantum states, such as conflicting electron spins or nuclei decaying and not-decaying, reduces to one definite state. We can mention first that associated with Eugene Wigner, because it is perhaps the easiest to dismiss. According to Wigner, the intervention of the consciousness of the observer triggers a superposition of states of a system into some definite state. The observer looks and the cat either dies or lives. It is the entry of the measurement signal into the consciousness of the observer which triggers the decision as to which of the possible outcomes is observed. But, it will be said, surely the cat is alive or dead before the human being looks in, and independently of our looking in. Wigner, though, is insistent that consciousness can act on matter and regards mind–matter interaction as a real problem and one crucial for our understanding of the world. While one may sympathize with Wigner on these points, the idea that it is consciousness which brings about wave-packet collapses is terribly hard to stomach. It is not simply that the mechanism whereby this is supposed to happen in the case of Schrödinger's cat is underdescribed. We are also, in Wigner's view, asked in effect to accept that 'in the thousands of millions of years before conscious life emerged in the world—and still today in those extensive parts of the universe where no conscious life has yet developed—no wavepacket has ever collapsed, no atom for certain decayed.'[18]

In contrast to Wigner's approach, the standard or Copenhagen interpretation of quantum mechanics does not make the collapse of the wave-packet or the decay of a nucleus dependent on conscious intervention. It says, rather, that the fuzziness of superposition is eliminated and things get made definite once the objects involved are sufficiently large. One particularly important class of large objects in this respect consists of our measuring instruments. We definitely read on an instrument that an electron spin is 'up' or 'down'; we definitely hear the click of a Geiger counter. Before that, though, all we can say about the underlying system is that we have a formula which simply assigns a statistical probability to a given system being in a particular state. The formula, on this view, is to be read as saying nothing about an individual state, but speaks only of the ratio of decay to non-decay, for

[18] J. Polkinghorne, *The Quantum World* (Penguin Books, Harmondsworth, 1986), p. 66.

example, in a collection of radioactive particles. The super-position of states is not taken literally, and the fuzziness is to be regarded as a function of our theories rather than of reality itself.

There is a certain undeniable attractiveness in avoiding a literal interpretation of superposition, but this does not explain how or why the larger system is something which can be definitely known, while the smaller systems underlying it remain—at least from our point of view—forever indeterminate and unknowable with precision. Nor does it explain how the actual decay of a nucleus and the effects of such decay, as in the case of Schrödinger's cat, is related to the probabilistic formula predicting a 50 per cent chance of decay of the nucleus within the allotted time-span. In considering the Schrödinger case in which there is said to be a 50 per cent chance of a nucleus decaying within one hour, either the probabilistic formula does reflect the reality, in which case nothing actually determines the outcome of the fundamentally random process, and we are left with the job of explaining why there is decay as opposed to non-decay in any particular nucleus. Alternatively, we say, as the Copenhagen school tended to, that there is a fundamental barrier to our knowledge of the objective physical world, which implies that some parts of it are forever unknowable, and that the best we can do is to produce probabilistic and fuzzy descriptions of it.

This fuzziness does not operate only at the level of uncertainty about what exact state a given particle is in. There is also the well-known problem about not being able to ascertain simultaneously the exact position and momentum of sub-atomic particles, despite being able to find out either separately. And there is the two-slit experiment regarding the emission of sub-atomic particles through a screen with two slits in it and their resulting distribution on a further screen. It appears that even though a group of particles goes through one of the slits, their distribution on the further screen is affected by the mere fact of the second slit being open even though they do not go through it or near it. As this is what would be expected if waves were being produced, going through both slits, and interfering with each other behind the slits, quantum theorists speak of the particles in question being both waves and particles, of 'wave–particle duality' as it is known. Niels Bohr summed up the problems of the simultaneous measurement of position and momentum of particles, and of the wave–particle duality, in what he called the principle of complementarity. On Bohr's attitude to this principle, Karl Popper wrote

when he accepted quantum mechanics as the end of the road, it was partly in despair; only classical physics was understandable, was a description of reality. Quantum mechanics was not a description of reality. Such a description was impossible to achieve in the atomic region; apparently because no such reality existed: the understandable reality ended where classical physics ended. The nearest to an understanding of atoms was his own principle of complementarity.[19]

But this leaves quite unexplained how a non-description of a non-reality can relate so successfully to actual descriptions of and manipulations of the real world (deaths of cats, explodings of bombs, production of electricity, and so on). Compared to Bohr's approach—which in effect attributes the fuzziness of our knowledge to the fuzziness of the world itself at the level in question—we might well be drawn to thinking of the quantum descriptions we have as essentially incomplete descriptions of the real world.

The idea that there is something radically incomplete in quantum theory is essentially what is involved in the so-called 'hidden-variable' approach, associated in recent times with David Bohm in particular. The intuition here is that whenever quantum theory postulates an indeterminacy or

[19] K. R. Popper, *Quantum Theory and the Schism in Physics* (Hutchinson, London, 1982), pp. 9–10.

describes a system in merely probabilistic terms, we are to conclude that if we knew more about the situation we would see all was perfectly determinate. Associated with phenomena of light, for example, there are both particles (which can be seen) and associated waves (which are inferred as creating the interference patterns we get in the two-slit experiment). In the Schrödinger cat case, presumably, in the system consisting of the radioactive material, there are (as yet unknown) forces or particles which determine whether or not a given nucleus will decay. The difficulties we have with radioactive decay which lead us to think in terms of superpositions of conflicting states are due to our ignorance of the complete situation.

Hidden-variable accounts are clearly agreeable to common sense, as they were to Einstein, who strongly disliked the Copenhagen interpretation of quantum physics, and its implication that random states of affairs are responsible for the world we can observe and act in. The quantum world is, on the hidden-variable view, perfectly determinate and definite, and there is no theoretical problem about Schrödinger's cat, or the transition from small to large; no problem, that is, beyond our ignorance of the totality of factors operative in the small. While not wishing to suggest that the question is closed, it must be admitted that hidden-variable accounts have not generally found favour with physicists. Bohm is credited with having produced a hidden-variable theory which reproduces all the empirical predictions of quantum mechanics, thus refuting the long-held belief that there could be no such theory. None the less, Bohm's theory is, according to Polkinghorne, 'pretty weird',[20] introducing, for example, an unobservable and hitherto unsuspected wave which comes into play in the two-slit experiment when both slits are open, just so as to produce the interference pattern manifested by particles going through the slits. Wave–particle duality is thus

[20] Polkinghorne, *The Quantum World*, p. 57.

severed, but it must be admitted that the explanation is *ad hoc*. The wave, along with other hidden variables, is introduced just to explain a mysterious quantum-mechanical effect. Unless and until these postulated hidden variables are independently identified, the suspicion must remain that they have no real existence, any more than the intra-Mercurial planet Vulcan had, and that quantum mechanics must remain as it standardly is, with its fuzziness, dualities, and complementarities.

Assuming that there are no hidden variables and unless we interpret quantum mechanics in a completely subjectivistic way, as not speaking about a real world at all and hence not to be interpreted in any literal way, the problem of Schrödinger's cat remains. We have to explain how a system in no single definite state gets pushed into a definite state. A further approach to this problem which we have not considered so far, largely on grounds of its metaphysical implausibility, is the many-worlds hypothesis of Everett, Wheeler, and Graham, according to which in effect everything possible in quantum mechanics actually happens in some world or other. Thus if in this world Schrödinger's cat lives, in some other world it or some replica of it dies. When we speak of the superposition of an ensemble of states of a system, we are really to be thought of as referring to each of the many worlds in which one or other of those states is actual. Apart from its intrinsic implausibility this proposal would not really help us with Schrödinger's cat. It would not explain why, in the world in which this 'I' is writing this now, the cat lives (or dies). I do not think it would be helpful to say (as no doubt would be said) that for every world in which I am writing this now and the cat lives, there is some other world in which a replica of me is writing the same stuff and the (or some other?) cat dies. I am interested in this world, in which I am now, and in why the cat dies or lives here, and not in the behaviour of replicas of myself or the cat in other worlds. So we are still left with Schrödinger's problem, the problem of integrating one level of description of the world with another. How are we to relate a level of description of reality in which there is fuzziness, chance, and indeterminacy with one which is quite definite? Or do we accept the conclusion that for some reason quantum-mechanical descriptions never apply literally to single events or situations, but only to the probabilities of given outcomes occurring in groups of such events or situations? If so, why should these be such an impediment to knowledge of the individual system?

In this section, I have been mainly interested in illustrating that difficulties can and do arise in integrating different parts of our physical picture of the world, or, perhaps better, in integrating our pictures of different parts of the physical world. Quantum mechanics, with its assumption of superpositions of states of given system, and classical mechanics, with its definiteness and lack of fuzziness would seem to be a good illustration of the sorts of problems that can arise here. Similar problems would also arise if, as I have hinted, the laws we have discovered turn out to apply only to some parts of space and time. What happens at the borders, where different conditions might apply? I do not pretend to answer this question, any more than I would presume to solve the problem of transition from indefiniteness to definiteness posed by Schrödinger's cat. But the problem posed by the cat should certainly make us wary of thinking that we are close to an absolute picture of the world, in which all the elements mesh smoothly and seamlessly.

7
Probability

As will be obvious from the last section of Chapter 6, the notion of probability plays a dominant role in current physical theory, and so some qualification of the straightforward relationship we have so far been assuming between prediction and explanation is now called for. We said in Chapter 1 that the deduction of predictions of specific events was characteristic of scientific explanations and, more loosely, that we expect scientific explanations to have predictive power. We must now show the way in which probabilistic theories may be seen to fulfil these basic criteria for scientific explanations.

Probabilistic Explanations

Jones smokes all his adult life and dies, at the age of 55, of lung cancer. We have a well-established theory to the effect that smoking greatly increases the risk of death from lung cancer and related diseases. An autopsy shows that Jones's lungs are indeed caked with smoke particles and his blood permeated with nicotine. We surely have a good explanation here of Jones's death. Yet Smith, who also smokes all his adult life, dies at the ripe old age of 95, by stepping under a bus, while still in the rudest of rude health. But we do not take this piece of evidence to be any sort of a counter-example to our theory about smoking and cancer, nor would we necessarily be perturbed by 50 cases of Smiths, or 500, or even 5,000. This is because our theory states that smoking *tends* to cause lung cancer, or makes lung cancer *probable*, and such theories are not refuted by a sample of cases in which there is smoking without lung cancer. On the other hand, were our theory of the universal form, '*all* cases of smoking cause lung cancer', then just one counter-example would be troubling, let alone 50 or 500 or 5,000.

The first moral to draw from this story is that what we will call probabilistic theories do not lead to straightforward predictions of single, isolated events. Smoking making cancer even highly probable does not entail that Smith (or Jones) will contract cancer, whereas Newton's laws do make specific predictions about the behaviour of (say) individual planets. We can say exactly what Uranus's behaviour or Mercury's behaviour should be like on Newtonian principles, given a complete description of the relevant factors. On the other hand, however complete a description we have of Smith or Jones and their circumstances, from a probabilistic theory we will not be able to deduce any prediction about their contracting or failing to contract cancer.

To see this point more clearly, let us assume that our theory does not just talk vaguely about likelihoods or probabilities, but actually assigns a specific numerical probability to a specific type of outcome. And to make this more concrete, let us take the theory about the fall of a fair die, which says (as would be expected) that there is a 5 in 6 chance of throwing some number other than a six. If we call this outcome 'a', and 'b' the action of throwing the die, we can write our theory (T) as

$$P(a, b) = 5/6$$

(i.e. the probability of a given b is 5 in 6). But, even though T assigns a very high probability to a, it is clear that the next throw could be a six, and so could the next two throws, or even the next three or four or five throws. T in fact says nothing about any specific throws, or even about runs of throws. A run of five consecutive sixes, for example, is not ruled out by T, provided that, over a long run of throws, they are balanced out by twenty-five non-sixes. A theory like T is about distributions of properties in classes of events considered collectively, rather than about single events within those classes. Or, to save the spirit of what we said earlier about scientific theories characteristically producing predictions of specific events, we will have to say that probabilistic theories make predictions of specific events regarded collectively (as members of classes of events), rather than predictions of those events taken singly and in isolation.

What, though, about the predictive power of probabilistic theories? Here we reach a further crucial distinction between universal and probabilistic theories, for while we can see predicted effects as being deducible from universal theories, no predictions are strictly deducible from a probabilistic theory. In the case of our T, even a very long run, of say 300 consecutive sixes, is not strictly ruled out. Such a run, however unlikely if T were true, could in the end be balanced out if thousands or millions of throws were made with the die. Any sequence of throws is compatible with the truth of T, on the assumption that in the long run the proportion of sixes to other throws turns out to be 1 in 6. And the problem that is raised for the testing and falsifying of T is that in T we are given no indication of how long that long run might have to be. It is not even the existence of an actual long run that militates against T's falsification by a disproportionate collection of sixes. The mere possibility that a long run of non-sixes might balance an imbalance of sixes prevents us from declaring T falsified as it stands.

In practice, if we got 300 consecutive sixes we would not persevere with T. Ten consecutive sixes would probably be enough to lead most people to question the fairness of the die; 30 in succession would be pretty clear evidence that we had a loaded die. And these obvious facts of experience are recognized in the way scientists and statisticians actually treat probabilistic theories. In the case of our T, although

300 consecutive sixes is not ruled out logically, we can say that, on the assumption that each successive throw is not dependent on the previous throws, such a run is statistically highly unlikely. If T is true, the likelihood in any sequence of throws that the proportion of sixes to non-sixes actually approaches that stated by T, itself approaches a probability of one. (Except in the rather theoretical case of random selections from infinite sets, assigning a probability of one to an event is tantamount to saying that it will definitely happen; a probability of zero means the event will not happen; intermediate cases are assigned probability values between zero and one.)

From what has just been said, it follows that in sequences of 1,000 throws, given that T is true, it is statistically very unlikely that more than a tiny proportion of sequences will deviate markedly from the ratio of sixes to non-sixes postulated by T, and the likelihoods of specific deviations can be computed. To get even one highly unlikely sequence can then be regarded as throwing doubt on the truth of T, and several highly unlikely sequences can be regarded as, practically speaking, sufficient to lead us to reject T. If, for example, we are dealing with coin-tosses and the probability of throwing heads is actually 0.5, after 40,000 tosses, the chances are 999 to 1 that our actual frequency will be within 1 per cent of 0.5 (i.e. $0.5 \pm .01$). So even though probabilistic theories cannot logically be refuted (for, it must be stressed, what is even 99.9 per cent unlikely *could* happen) implications regarding statistically unlikely outcomes given their truth can be sufficient to provide reasonable grounds for their empirical testing and, if need be, for their rejection as well.

The mathematical basis of this form of testing of probabilistic theories is Bernouilli's theorem, the so-called Law of Large Numbers. This theorem states in effect that any sufficiently large sample drawn from a larger population is likely to match the parent population in its distribution of various characteristics. As the size of the sample increases, this likelihood approaches certainty (P = 1). Returning to the testing of probabilistic theories, we then have some kind of assurance that if a theory of ours says that a certain characteristic is distributed throughout a given population according to a certain ratio, large samples taken from that population are unlikely to deviate much from that ratio in their possession of the given characteristic. If the ratio of sixes in fair dice throws is 1 in 6, then a large number of throws with a fair die is likely to show a ratio close to 1 in 6 between sixes and other throws.

Now while all this is mathematically perfectly sound, its soundness as a reason for actually rejecting a hypothesis rests on the tacit assumption that in our sample populations we are dealing with typical samples, or with samples that have as much chance of being picked as any others. We would be justified in rejecting some hypothesis about the distribution of the property we are interested in, on the basis of some sufficiently large sample or samples deviating from the distribution, only if we had no reason to suppose that these samples were atypical. As A. J. Ayer comments, 'it is easy to show that we can reasonably assume this if we are entitled to make the assumption that any one sample of a given size is as likely to be selected as any other'.[1] Ayer goes on to comment that this assumption is innocent enough if we equate the probability of getting a particular sample with the ratio it bears to the total number of possible selections, and assuming that we had an equal chance of picking on any particular sample, but in empirical testing we do not have an equal chance of getting any particular sample. As Ayer says, 'our samples are drawn from a tiny selection of the universe during a very short period of time', and without illicitly assuming some principle to the effect that nature is uniform we cannot legitimately generalize from them. We have, in

[1] A. J. Ayer, *Probability and Evidence* (Macmillan, London, 1972), p. 41.

other words, no a priori reason for thinking that they might not be deviant samples and very unlikely in the universe as a whole.

The problem we are alluding to here is, of course, the general problem of induction; we are considering the probabilistic analogue of the problem we encountered earlier when establishing universal theories on the basis of data we collected simply in our region of space and time. It is true that the probabilistic problem refers to the refutation of theories as well as to their confirmation, in that deviant samples collected from a small range of all possible samples ought not necessarily to count *against* probabilistic hypotheses. None the less, what we are primarily interested in is our region of space and time and the conditions obtaining there, and the reliability of our hypotheses relative to those conditions. Just as we use inductive generalizations from our experience to base our non-probabilistic hypotheses about regularities local to us, so unless we know of positive reasons to the contrary we should continue to assume that the samples we actually get of characteristics in populations are representative of larger populations, at least as far as our region of space and time is concerned, in order to test our probabilistic hypotheses. So long as we do not think that this use of the Law of Large Numbers establishes anything for the universe as a whole or provides strict proofs regarding our future, we can still legitimately use it to eliminate probabilistic hypotheses which do not appear to be true of the conditions relating to our region of space and time.

So, probabilistic theories can be regarded as both empirically significant and empirically testable. None the less, it might be felt that there is still a sense in which they are less than fully explanatory. If we take Newton's theory as an example, when people were given the formulae relevant to determining the courses of the planets and they found the planets obeying the formulae, they felt they had unearthed the principles on which the planets were operating. This feeling is understandable because of the belief that these formulae applied without exceptions to all planetary movement, and to a host of other phenomena as well. By contrast, smoking cannot be seen as the complete explanation of the death of our imaginary Jones, precisely because, as the example of Smith showed, there is no exceptionless principle linking smoking with cancer. In the case of Jones's death, it would be quite natural to think that his smoking was only part of the explanation, and that a full account of what caused his death would turn up factors relevant in his case which were absent in the case of Smith. It might be that Jones had some as yet unknown difference in his genetic make-up, which made him more prone to cancer than Smith. Smith, lacking this susceptibility, was perhaps immune to the carcinogenic properties of nicotine.

Consideration of the contrast between the cases of planetary motion and cancer prompts the following thought— which is laid out explicitly in Davidson's paper 'Causal Relations'.[2] When we speak of something like Jones's death being caused by smoking, where the relation between cause and effect is probabilistic rather than deterministic, we should think of our description of the cause as essentially incomplete, and relying implicitly on the possibility of there being a complete description of the cause. Thus, while it is true that smoking did cause Jones's death by cancer, it did so only as a case of a person of a particular genetic constitution smoking in particular circumstances. Our incomplete description of the cause of his death thus rests on some fuller description, a fuller description which, we may suppose, would fall under some exceptionless law, in this case to the effect that people of such and such a constitution who smoke in such and such circumstances invariably contract cancer. Of course, we may not be able to specify the relevant constitution and circumstances, but the belief that they exist

[2] In D. Davidson, *Essays on Actions and Events* (Clarendon Press, Oxford, 1980), ch. 7.

shows that there need be no tension between our using probabilistic theories and the world being fully determined (that is, operating on the basis of exceptionless regularities).

Our use of probabilistic theories may be due to our ignorance of these further regularities and their specification (as presumably is the case in the cancer example), or it may be because giving a full specification of the further relevant factors and laws is of no interest to us, or, for some reason, not worth the effort involved. To an insurance company, for example, what is of interest is the distribution of deaths within specific categories of person, rather than the detailed causes of specific deaths. Collecting the data on which such specific predictions might be made would almost certainly be far too onerous and costly for the task in hand, assuming it could be done at all. If the insurance company gets the overall distribution of deaths within the relevant population correct, it will still make profits through fixing premiums so as to exceed pay-outs, and its clients and employees will be spared the doubtless costly and probably distasteful exercise of enquiring in detail into the lives and histories of the clients.

Davidson's analysis of probabilistic causes assumes that what is at issue here are incomplete ways of describing fully determined situations. For many situations, this may well be correct. The cancer example may be a case in point, and it is quite natural to feel that cases of throwing dice are actually fully determined as well. If we knew all the details of the hand movements, air currents etc., and the relevant laws, we would be able to see why—in terms of exceptionless laws— this particular throw resulted in a five rather than some other number. Indeed, to speak of what we are throwing as a fair die is to say no more than that it is so formed as to have an equal chance of landing on any one of six faces, and while this no doubt explains something about the way the die lands on a particular occasion, it clearly cannot explain why it lands on one face rather than on one of the others. But we cannot infer that there are not other and full explanations of that, though once again applying the relevant theories in practice to particular throws would be a herculean task.

However, while some uses of probabilistic theories may not preclude the possibility of full, deterministic explanations of the events in question which bring them under exceptionless laws, as we have seen in the case of quantum theory, modern physics is apparently content in some areas to work with probabilistic theories without any assumption that there may be fuller explanations in the offing. Bohm's postulation of hidden variables underlying probabilistic effects in the quantum world, which has obvious parallels with the Davidsonian analysis of probabilistic causes, has not, as we have seen, found general favour with physicists. We are told, for example, that radioactive substances have certain half-lives. This means that over a given period of time (the half-life), half the atomic nuclei of a quantity of the substance will have decayed. As in the example of the theory about the die, there is no explanation offered here of why particular nuclei do or don't decay within that time. All we are told is that a certain *proportion* will decay in the period, just as our die theory says simply that one-sixth of the throws will be sixes. In neither case is there any attempt to explain or predict particular outcomes. But with atomic half-lives, unlike in the die case, we are asked to accept that no further explanation of the behaviour of a particular nucleus could be given. Its decaying or not decaying is a matter of chance, something unpredictable and undetermined, and not due to its age or to any other of its properties. All that is determined is that members of sequences of atomic nuclei will decay in line with their half-life.

In what sense, then, do we have an explanation of a particular instance of radioactive decay, when we are told, for example, that the half-life of carbon 14 is 5730 ± 30 years? We are being told that, after this period, half of a given isotope of carbon 14 (or radio-carbon) will have become nitrogen, owing to the emission of a beta particle or electron (which, incidentally, is what carbon 14 is formed from, when cosmic rays strike nitrogen in the atmosphere). But in this talk of half-lives and the surrounding theory, we are told nothing about the behaviour or decay of the particular nuclei of a radio-carbon isotope. There is a clear sense in which their behaviour is left unexplained; and, unlike the case of die-throws, something here is regarded as fundamentally inexplicable. The use of probabilistic theories in quantum theory goes hand in hand with a belief that in certain respects the universe is not determined; that determinism is false and indeterminism true.

Against this background, one might once again push the point about the explanatory power of probabilistic theories. In themselves, they do not give a full explanation of individual events. In some cases, if indeterminism is true, there may indeed be no full explanation of individual events, and no theory which would allow us to predict the future behaviour of individual particles. Probabilistic theories explain effects, in so far as they do, by pointing out statistical tendencies within populations. Their universality and their testability comes at the level of populations and samples of populations. Any sample of radio-carbon will be held to have a half-life of 5730 ± 30 years. Any group of photons fired at a screen will have such and such a distribution, and so on. But below the level of tendencies within populations we find random and fundamentally inexplicable behaviour, and this is reflected in the use of probabilistic theories in these areas. Another way of looking at the sub-atomic world would be more holistically, rather than thinking of sub-atomic particles as behaving in independence of their environments. While doing this may well be right, and may give some reason for the difficulty of prediction in the sub-atomic realm, a more holistic approach would not of itself answer the problems of predicting the behaviour of individual particles. It would not enable us to predict which nuclei in a sample of radio-carbon will decay or explain why particular particles in the two-slit experiment take one course rather than another.

It may, of course, be hard to establish examples of tendencies within populations, when actual populations are mixed in relevant respects. There may be a tendency for smokers to contract heart disease, but, as Nancy Cartwright points out, if all or most smokers counteract this tendency by being good exercisers, the incidence of heart disease among smokers may actually be less than among the population as a whole.[3] Nevertheless, in speaking of a causal link between smoking and heart disease, or between smoking and cancer, as we did earlier, we will be implying that in populations where there are no countervailing factors, smokers will be more inclined to get heart disease than non-smokers. This reference to the absence of countervailing factors parallels the similar qualification, in testing universal laws, concerning the absence of interfering factors. But, in the probabilistic case, our explanations and understanding of causality will be in terms of well-defined mathematical tendencies within populations, rather than the constant conjunctions of deterministic causes and universal laws.

In the case of probabilistic laws, our explanations will be explanatory to the extent that seeing occurrences in terms of tendencies within populations explains outcomes of particular sorts. This will be true in the type of case where a given tendency within a population makes some outcome more probable than it would otherwise have been (and we then think in terms of some causal link between the feature of the population, the tendency, and the outcome, as in the case of smoking and heart disease). But with a probabilistic theory, there will also be a type of explanation given of why the less probable outcomes occur. If the explanation of someone's cancer or heart disease by their smoking rests simply on a

[3] Cf. Nancy Cartwright, *How the Laws of Physics Lie* (Clarendon Press, Oxford, 1983), pp. 23–4.

statistical distribution, then, provided cases of smokers lacking of heart disease or cancer fall within the appropriate statistical distribution, there will be a sense in which their not getting the disease is also explained by the fact that they belong to a population in which there is a tendency for a certain proportion of its members to turn out some way. To say that there is a *tendency* within a population is to say that some members of that population will *not* have or do whatever the tendency is to have or do. We naturally feel in many cases that there must be underlying reasons as to why some members of the population do contract the disease, or whatever, and others do not. But in genuinely indeterministic cases, there are no further reasons, and all we can point to are tendencies within populations—without being able to explain why some members of that population turn out one way and others another. In such cases, our search for further causal explanations of particular outcomes would be doomed to failure.

Interpretations of Probability

In the previous section, we considered the nature of probabilistic theories in science, their testing, and the extent to which they could be seen as providing explanations of empirical events. In this section we will consider some of the ways in which philosophers have understood statements of probability. In doing so we will be looking at some of the ground already covered from a different angle, and we will also draw on some of our remarks on quantum theory in Chapter 6. In doing so, we should be able to get a little clearer about some of the issues relating to both probability and indeterminism.

In giving a philosophical account of probability which will be relevant to the way notions of probability are used in scientific contexts, we will be attempting to give an

interpretation of the mathematically formalized calculus of
probability or chances. There is widespread agreement as to
the mathematical structure of this calculus, which enables us
to assign numerical degrees of probability to events or
hypotheses. But agreement on the basic axioms and rules for
assigning numerical degrees of probability does not fix the
interpretation of probability itself; it does not tell us what is
meant by assigning something a given degree of probability.
More specifically, what we are here interested in is giving an
interpretation to statements of the form

P (a, b) = r

where P (a, b) is the probability of a given b and r is some
number between 0 and 1 (these limits included). Philosophical accounts of probability can be broadly divided into
subjective accounts, which regard probability statements in
terms of what we are entitled to believe on given evidence,
and objective accounts, which interpret probability statements as referring directly to tendencies of various sorts
existing in the real world.

The classical theory of probability, usually associated with
Laplace's *Philosophical Essay on Probability* of 1814, can be
seen as a subjective theory, in that it envisages judgements of
probability as being applicable primarily in cases where we
have no more reason to expect one outcome of a given type as
opposed to some other outcome of that type. Thus, we would
have no more reason to expect a coin to land heads than tails,
and we can speak of the probability of its landing heads (or
tails) as 1 in 2. In general the classical theory will analyse the
probability of one type of (equiprobable) alternative outcome
of some form as being the ratio of favourable cases to the
ratio of all possible cases, 1 in 2 in the case of coin-tosses
resulting in heads, 1 in 6 in the case of die falling on six, 1 in
36 in the case of throwing double six with two dice, and so
on.

Laplace and his followers would typically calculate this

ratio on a priori grounds, in terms of a relationship between given information and the hypothesis in question. In the case of dice or coins, the information amounts in fact to a lack of information or ignorance about any particular fall. Any particular fall is treated as being as probable (or improbable) as any other. There would be no reference in such calculations to observed frequencies or ratios of falls or to other matters of fact, such as the physical constitution of a die or coin. What we would have here is simply an application of the mathematical calculus of chances (or probability calculus), and this immediately raises the problem of explaining how a piece of abstract mathematics can properly be applied to real games of coin-tossing, backgammon, and the like.

It is clear that we could not use results achieved in a purely a priori way, if there was not in our actual set-up an equal probability of any particular outcome. If 200 red and 200 white balls in an urn were arranged so that all the red balls were placed on top of the white ones, there would not be a 1 in 2 chance of drawing a white ball in the first draw. Hence the requirement, in the classical conception, that the basic alternatives be equiprobable. At this point, though, in any actual application of the calculus of chances, we will be going beyond pure mathematics, and making empirical assumptions to the effect that a particular die or coin is fair, or that balls in an urn are randomly distributed. In other words, although like a backgammon player we can calculate a priori what the chances of various outcomes will be, given that we have a certain number of equiprobable basic alternative outcomes, one of which we have no more reason to expect than any other, whether our system does, like backgammon dice, constitute such a set-up is not an a priori matter at all, but is based on empirical facts and requires evidence of equiprobable runs.

The classical Laplacian conception of probability assumes that the basic alternatives being assessed are equiprobable. This immediately limits its application for scientific purposes, for as Hempel points out[4] in science we frequently deal with cases where the basic alternatives are not regarded as equiprobable. Examples include the sorts of case we have already considered, such as the step-by-step decay of the atoms of radioactive substances and the transition of particles from one state or location to another. Even in gambling, where we were dealing with biased coins or dice and hence with unequal alternatives, the classical conception would not be applicable. Its application then is going to be highly restricted.

In addition to problems over the applicability of a theory based on the assumption of equiprobable alternatives, there is a further significant problem with the classical account of probability. Even on the assumption of equiprobability or equal ignorance regarding various outcomes, the same outcome can be assigned different probabilities depending on how it is described. We have a pack of four cards, consisting of two red and two black cards, and we want to compute the probability of drawing a hand of two cards of the same colour from our pack. If we take our equiprobable basic alternatives to be the individual cards we draw, there will be six possible hands, of which only two will be of the same colour, so our probability comes out as 2 in 6. On the other hand, if we take the colour make-up of the completed hands as the basic alternatives, there are only three possible outcomes, two of which again will be of the same colour, and so our probability comes out as 2 in 3. It is not just that it is unsettling to get the same event emerging with different probabilities; the basic problem is that the notion of parity of reasons as between alternatives gives us no guide for the selection of our basic vocabulary or for one of the two probabilities. For guidance here we would presumably have to go to actual drawings of hands, sullying the purely mathematical calculation of chances with empirical observation.

[4] Cf. *The Philosophy of Natural Science* (Prentice Hall, Englewood Cliffs, NJ, 1966), p. 61.

Difficulties analogous to the one just encountered would also appear to attend any purely logical interpretation of probability statements, such as that favoured by Carnap. Logical interpretations of probability operate on the assumption that the probability of a hypothesis is the extent to which the evidence for it logically supports it. Probability is thus seen as a logical relation, whose limits are logical implication of hypothesis by evidence (P = 1) or logical contradiction of hypothesis by evidence (P = 0). The intermediate probabilities are calculated in complex ways which derive from the notion of logical space in Wittgenstein's *Tractatus Logico-Philosophicus*. In one of Carnap's systems, for example, we take all the basic predicates of our language and the names of all the individuals in the universe we are considering, and we then construct what are known as 'state descriptions', statements in which every name is assigned one out of each primitive predicate and its negation. (If our language has just one such predicate, 'green', then a state description will be a conjunction of sentences, saying of each and every individual that it is green or that it is not green.) Each state description constitutes a possible world or universe, relative to the language; in terms of that language it completely describes our universe. Each state description is initially assigned some positive degree of probability, and on this basis, any sentence of the language can be assigned its initial probability (the sum of all the initial degrees of probability of the state descriptions in which it features). By means of the calculus of chances, we can then ascertain the degree of probability, or support, given to a sentence which expresses a hypothesis we are testing, by the sentence which expresses the evidence we have collected in favour of that hypothesis; we can thus calculate what, on given evidence, the probability of given events will be.

For empirical purposes, unfortunately, the purity and applicability of this approach to the probability of actual hypotheses and of actual events is vitiated not only by the inductivism implicit in thinking that our evidence is at any given time representative of all possible evidence, but more specifically by the fact that our assessments of probability will depend both on our initial weighting of the state descriptions and of our choice of language. The choice of language affects probability assessments for reasons analogous to those we saw in the case of the card decks, where all depends on whether we take the individual cards or whole hands to be the basic individuals in the language. The former factor is relevant because once it is allowed that we cannot assume that all state descriptions are equally probable or improbable (because that itself is a substantive assumption of any actual universe), we are left with little guidance on what the weightings ought to be.

The classical and logical theories of probability are not the only examples of subjective approaches to probability, if by 'subjective' we mean that the probability of a statement or of an event described by that statement is equivalent to the degree to which one might be justified, on a given body of evidence, in believing that the statement is true or that the event will happen. There are, indeed, subjective accounts of probability which do not rest on assumptions of equiprobability or Carnapian manipulations of state descriptions. In order to bring out as starkly as possible the difference between subjective and objective accounts of probability, we may take the essence of subjective theories to be simply the belief that, in the words of Bruno de Finetti (a leading subjectivist about probability), a statement of probability does not reflect anything 'rational, positive or metaphysical' in the world;[5] it is a merely psychological device which we use when we are in ignorance of the full facts of a situation. Saying that the next toss of a coin has a 1 in 2 probability of being heads is simply an expression of our subjective belief

[5] In 'Foresight: Its Logical Laws, Its Subjective Sources', in H. Kyburg and H. Smokler (eds.), *Studies in Subjective Philosophy* (Wiley, New York, 1964), pp. 97–158, at p. 152.

based on ignorance of the full facts of the situation (which, we assume, would if we knew them allow us to make a definite prediction).

Now, it is certainly true that belief is, or ought to be, a function of evidence. And it is also true, as subjectivists such as de Finetti point out, that increasing evidence will often tend to modify initially quite divergent beliefs on factual matters. Interpreting probability in terms of subjective belief can, then, seem reasonable when we are talking about the probability of hypotheses given bits of evidence, and, as de Finetti and others show, the probability calculus in general and Bayes's theorem in particular can give a reasonable account of the way rational belief in a hypothesis strengthens or weakens on increasing evidence. But, as objectivists about probability will urge, when we talk about the probability of events in the physical world, we are not fundamentally talking about the degree of support some evidence gives to a hypothesis or about the extent to which it might be partially entailed by some evidence in a particular language. These latter notions rest on what we are fundamentally referring to, which are tendencies in the real world, and treating probability statements in science as quasi-logical entailments from some evidence or as degrees of subjective belief completely fails to account either for their application to the real world or for the way they are so often brilliantly confirmed in the real world. We could starkly put to the subjectivist Popper's often-repeated demand for an explanation of the manner in which what the subjectivist takes as tantamount to an expression of ignorance is so often tested *and* corroborated.

The feeling that the notion of probability as referring to real tendencies in the world must be more relevant (in scientific contexts at least) than probability as subjective degree of belief is nicely supported by what Popper refers to as the paradox of ideal evidence.[6] Suppose, in line with

[6] In *The Logic of Scientific Discovery*, pp. 407–9.

subjectivism, we think of a statement that some arbitrary coin-toss has a 1 in 2 probability of being heads as primarily a statement of ignorance. ('I am 50 per cent ignorant about the outcome of the toss.') We then observe a very large number of tosses with the coin, at the end of which we still hold that the next toss has a 1 in 2 probability of being heads. If, as the subjectivist holds, probability statements reflect our ignorance about conditions and outcomes of individual events, we are still as ignorant as we were when we started, and our long run of tosses seems to have taught us nothing. Against this interpretation, though, surely the natural thing is to say that our tossing has taught us something quite real about a tendency that particular coin really has to fall heads or tails indifferently, and our observations are to be seen as (in this case) confirming the hypothesis that it really has such a tendency.

Objectivist accounts of probability see probability statements as referring to real tendencies individuals or sequences have to manifest certain patterns of outcome. In the case of the long sequence of coin-tosses, we can then regard our observations as giving some rational backing to the hypothesis that the coin has a 1 in 2 tendency to land heads. As far as our state of knowledge goes, we know more at the end of the tossing sequence than we knew at the beginning, more, that is to say, about the truth of the hypothesis that the coin has the tendency in question. We can say this because we are here interpreting the statement

$$P(a, b) = 0.5$$

(where b is a coin-toss, and a the coin landing heads) as referring to some tendency the coin or sequences of coin-tosses has, rather than to the extent of our knowledge or ignorance regarding coin tosses.

Treating probability statements objectively in this sort of way, though, does not tell us exactly how they are to be understood. According to what is known as the frequency

view of probability, in saying that there is a 1 in 2 probability
of the next toss of the coin being heads, I am not really
speaking about the next toss—which, after all, will be definitely one thing or another, and may indeed be determined to
be one thing or another—but about a whole class of throws,
of which the next toss is only one element. My probability
statement is, on this frequency view, really about the relative
frequency of heads and tails in a whole sequence of tosses.
Apart from the difficulties involved in attributing probabilities to single events, to which we will return, the frequency
view is attractive because it links talk of probabilities with
their mode of discovery; it is certainly the case that we often
estimate and check probability assessments on the basis of
observed frequencies of the relevant characteristics in populations of individuals or events.

The idea that probability statements refer to relative
frequencies within particular classes corresponds well
enough to many of the actual uses of probability statements
in science and statistics. To say, for example, that a child
conceived by a mother of over 42 has a 1 in 100 chance of
having Down's syndrome is to say that the frequency of the
syndrome over the population of such children is 1 in 100.
Where the populations in question are finite there is clearly
no problem, in theory at least, in speaking of the relative
frequency of a given characteristic within a population. Very
often, indeed, populations we are interested in are quite
circumscribed. The frequency of Down's syndrome may
well vary with time and place and other factors, such as diet
and nutrition; what a mother in England wants to know is the
chance of her child having the condition, and for her a more
relevant statistic, than the incidence of the syndrome over
the whole world and at all times, will be its frequency in the
given population in England in the last part of the twentieth
century. Of course, in estimating this probability we will
have to extrapolate on the basis of a limited sample of cases
(from 1975 to 1988, say), but the extrapolation may not seem

too risky, given similarity of circumstance. In any case, the whole population concerned is strictly finite and circumscribed, so there is no theoretical difficulty in analysing probability in terms of the relative frequency of the syndrome within that population.

On the other hand, we often want to talk of probabilities in populations which are not circumscribed or even necessarily finite. In quantum theory, as we have seen, we speak of probabilities which pertain within populations of particles which we would not want to say in advance were finite. How is relative frequency to be thought of in such cases? Furthermore, we want to think of our probability statements as entailing counterfactual statements about what world would have happened, but did not, and about the behaviour of possible entities; once again, it seems we have to envisage our frequencies operating over at least potentially infinite populations.

To address these difficulties, Richard von Mises, the architect of the modern frequency theory, sees probability in terms of the distribution of some property over a 'collective', a potentially infinite series of events in which a given property is distributed randomly. By insisting on the random occurrence of the property in question (by what he calls the axiom of randomness), von Mises intends to rule out sequences to which a gambler could successfully apply a gambling strategy. This is to ensure that any sub-sequences we encounter are genuinely chance-like, and representative of the whole. Von Mises also introduces what he calls the axiom of convergence; as we look at more and more cases in a sub-sequence of a collective or in the collective as a whole, the frequency of randomly distributed properties in either sub-sequence or collective tends to settle around some definite value, or, in von Mises's terms, reaches a frequency limit. The two axioms taken together have the result that as no sub-sequence we actually investigate is more ordered than the collective as a whole, as both are supposed to be equally random, the frequency limits of long enough sub-sequences will be similar to that of the collective as a whole. In this way, von Mises hopes to show how it is possible to infer derived distributions in derived collectives from given actual sequences; in so doing, he also presents a solution to the worrying problem of how to show that order and randomness are not inconsistent, and that by applying the calculus of chances we can indeed find order and regularity in randomness.

One obvious practical difficulty with von Mises's proposals is that of knowing whether we actually ever have a collective in his sense—i.e. a sequence which obeys either or both of his axioms. Von Mises's actual axiom of randomness requires that any sub-sequence we draw by any regular method from a collective should have the same frequency limit as any other. This is so strong a requirement that we must doubt whether it can be satisfied in practice. It is also the case that even though recorded frequencies cluster round a given value, this is never inconsistent with the series as a whole either eventually reaching a different limit or having no limit at all.[7] Partly to deal with these and other difficulties in von Mises's version of the frequency theory, Popper introduced a revised version, giving a different notion of randomness and one which ensured that no additional axiom of convergence was needed. In Popper's system, long enough sequences with high initial randomness are presumed, by the Law of Large Numbers, to be likely to reach the frequency limits of much longer sequences.[8] This does not, of course, get round the difficulty of knowing whether a highly random sequence we observe is actually representative of the population as a whole. The Law of Large Numbers may tell us that it would have to be highly deviant if it were, but that it is not highly deviant is not something we can be sure of, without something like a fair sampling postulate (in which case, as

---

[7] On both these points, cf. Ayer, *Probability and Evidence*, pp. 46–8.
[8] In *The Logic of Scientific Discovery*, pp. 154–91 and 359–62.

Ayer comments, talk of frequency limits is irrelevant). In Popper's proposal, there is also a problem arising from the demand that a sequence has high *initial* randomness, for it seems to imply that we must make some possibly erroneous and possibly uncheckable empirical assumptions about any observed regularity in, for example, the behaviour of a die we are throwing, as occurring long into its history, if we are to be allowed to apply the calculus of probability to it. Leaving this aside, though, the difficulty about deviant samples is only a version of the familiar problem of induction, and it is not clear how any theory of probability is going to get round that. There would, however, still be a major hurdle for the frequency theory to clear even if, not unreasonably for certain cases at least, we allowed its proponents something like a fair sampling postulate in practice, and were prepared to regard frequencies of characteristics in observed sub-sequences of larger populations as representative of those larger populations, whether infinite or not. The problem is that in many cases we want to attribute probabilities to single events and not simply to classes of events.

We can look at the problem of attributing probabilities to single events in a number of ways. We might begin by considering a die which is thrown only a few times or even not thrown at all before being destroyed. In such a case, there would be no actual range of cases to which some frequency value of particular outcomes could be attributed. Yet we would surely be inclined to say that the die had had various probabilities of being thrown with given outcomes, due to its structure. It is true that our *evidence* for saying this might well be that similar dice with similar structure had actually manifested frequency limits close to the values we want to attribute to the unthrown die, but it would not be unreasonable to distinguish our evidence (from runs of throws with similar die) from what it is evidence for. And what it is evidence for looks like some genuine physical tendency or propensity the unthrown die had, before it was destroyed, to manifest certain statistical regularities in throws. The throws were not, of course, actually undertaken, but put in this way the propensity, as we will call it, seems real enough, and a lot more real, in fact, that the infinite collectives of potential throws von Mises postulates as the subject of probability statements. Even with a die that had been thrown a lot, there can be no actual collective of throws. The infinite collective looks like some sort of theoretical construction arising from a propensity the die has to manifest certain frequency limits in actual sequences of throws.

The frequency theory cannot account for single happenings except in terms of theoretical classes to which those single happenings are presumed to belong. But the theory actually has to face an even more serious problem than the invocation of theoretical classes of events. Precisely because the frequency theory has to analyse probabilities relating to individual events or objects in terms of classes the individuals belong to, the probability to be ascribed to an individual object or event having a particular property will depend on the relative frequency of that property throughout the class the individual is seen as belonging to. But individuals can, of course, be seen as belonging to more than one class, and in cases where the frequency of the property is different in the different classes, the same individual will be ascribed more than one probability of having the same property.

We can illustrate this point by using a well-known example. We want to know the chances of Petersen, a Swede, being a Protestant. Suppose 95 per cent of Swedes are Protestants, so the odds are 19 to 1 on. But we also know that Petersen made a pilgrimage to Lourdes last year, and at least 95 per cent of such pilgrims are Catholics. So the odds are at least 19 to 1 against. We have a flat contradiction here, which can indeed be removed if we interpret the relevant probabilities as being relative frequencies. But then our saying that Petersen has a 19 to 1 chance of being a Protestant is simply a restatement of the fact that he is a Swede, while the 19 to 1 chance of his being a Catholic is just a restatement of the fact that he went to Lourdes.

We thus lose our original question about Petersen, and it is doubtful whether it can be resurrected within the compass of the frequency theory. Some may advise us to take the narrowest reference class to which individuals belong in order to estimate relevant probabilities in their case (Swedes at Lourdes, perhaps, in our example). But even if one could satisfactorily define narrowest reference class in this context, as Ayer points out, there can be no place for such advice within the terms of the frequency theory: 'one can attach no sense, within the limits of the theory, to saying that the choice of a narrower reference class yields a better estimate of my chances [of having a certain property]'.[9] The reason for this is that the frequency theory always reads statements about an individual's chances as an elliptical way of saying how some property is distributed through some reference class, and, provided the probabilities have been correctly estimated, there are no grounds within the frequency theory for preferring the choice of one reference class to another. It is just that the choice of different reference classes yields different information.

But we surely do want to think of probabilities of individuals having certain properties and, on occasion, we do think some reference classes provide more useful information than others for estimates of chances relating to the outcomes of individual events. Where the reference class in question can be seen in terms of the conditions which generate the outcomes involved, it might well seem natural to see probabilities in terms of actual propensities rather than in terms of frequencies. Or so, at least, it did to Popper when he moved from advocacy of a frequency theory of probability to what has been dubbed the propensity theory. He considers an example of a long series of throws with a loaded die,

[9] A. J. Ayer, *Probability and Evidence*, p. 52.

whose chance of landing on six is 1 in 4. Imagine two or at
most three throws with a fair die inserted into this series.
The chance of throwing a six in the whole series is still going
to be very close to 1 in 4. Nevertheless we will want to say of
the fair die throws that their chance of landing on six is 1 in 6,
and this even though in fact there are not enough actual
throws with the fair die on which to base this. The frequency
theorist will no doubt refer at this point to a virtual series of
throws with the fair die, whose probability of landing on six
approaches a frequency limit of 1 in 6. But—and this is the
crucial point for Popper—this virtual sequence is characterized, and indeed justified, in terms of the conditions which
generate it (structure of the fair die, random throwing, etc.).
The difference in probability of throwing a six as between
the long, loaded die sequence and the two or three fair die
throws is due entirely to difference in their generating
conditions.[10] Probability, then, is a property of the generating conditions of events; this is the basis of the propensity
theory of probability.

The propensity theory amounts to the claim that while
certain physical set-ups are random or unpredictable, as far
as their individual outcomes are concerned, repeated experiments or observations of the set-ups in question will show
statistical stability. This stability is seen as due to propensities inherent in the set-up, and these propensities are
regarded, by Popper at least, as actually existing but unobservable dispositional properties of the physical world. He
regards them as analogous to and an indeterministic generalization of Newtonian forces. What propensities bring about,
according to Popper, are not single events, but observed
frequencies in runs of events. Unlike deterministic forces,
they cannot bring about single events, for in a genuinely
indeterministic case *nothing* brings about the single event,
this nucleus decaying, the photon going *that* way. But still,

[10] The example is discussed in his *Realism and the Aim of Science* (Hutchinson, London, 1983), pp. 353–6.

according to Popper, the objective probability of a single event can and should be seen 'as a measure of an objective *propensity*—of the strength of the tendency, inherent in the specified physical situation, to realize the event—to make it happen'.[11]

Although Popper says that propensities are not Newtonian forces and only analogous to them, this quite natural-seeming talk about a propensity as the strength of a tendency in an individual event to make that event happen does make it look rather as if a propensity is a force, like gravity, say. But regarding propensities as quasi-forces makes it very hard to see how a strong propensity (or tendency)—say the 60 per cent tendency for a biased coin to land heads—could ever be overcome (as clearly it must be on occasion) by the corresponding weak (40 per cent) tendency for the same coin to land tails. When there are two competing forces, the stronger will always win. So clearly propensities cannot be forces. But this does not positively explain what they are. In particular, it does not explain how propensities are other than extrapolations from observed or conjectured frequencies in given cases. The empirically minded are likely to object here that talk of propensities is being invoked simply to explain the otherwise not further explicable distribution of frequencies in classes of happening. In saying that *this* carbon 14 nucleus has a 1 in 2 propensity to decay in 5730 ± 30 years, are we saying any more than that it belongs to a class of entities, half of which regularly decay over the period? Certainly talking of a propensity the single nucleus has to decay does not allow us to make any predictions about *its* longevity, about its decaying in the next instant or only after 10,000 years. And yet, the propensity theory is introduced to deal with the single event.

This objection is, of course, unfair. Propensities are not forces, nor does the theory claim to be able to produce predictions of a single event being such and such. None the

<hr>

[11] *Realism and the Aim of Science*, p. 395.

less, the onus is on the propensity theorist to show how his theory differs in practice from some kind of frequency theory, since propensities of single events on analysis seem to involve some sort of reference to frequencies in actual or potential runs of events.

Popper himself claims that the two-slit experiment in quantum physics provides evidence in favour of the propensity theory, showing that propensities are physically real.[12] He argues that just as inserting new pins on a pin-table changes the probabilities or propensities of balls rolling down the table, even when they don't actually go near the new pins, so opening the second slit changes the propensities of distribution of particles which actually go through the first slit. From the point of view of probability theory, what Popper says is correct. I insert a new pin down on the right-hand side of the pin-board, in such a way as to affect the course and final position of any ball which hits it. We can now say that the probability of any ball coming to some specific resting-place on the right-hand side of the board is altered, and this is true even of balls which actually go down the left-hand side (before they do so, of course). But obviously, in the pin-ball case, the actual courses of balls rolling down the left side of the board are not affected or interfered with by the mere insertion of a new pin on the right-hand side. In the two-slit experiment it is not merely change of the probability of a particle reaching a given point which is altered by opening the second slit: it is its actual course. Physical interference of this sort is quite different from the change of probability in the pin-ball case, and it is hard to see how invoking the propensity theory throws any light on it. Indeed, in wishing to analyse the two-slit experiment in terms of probabilistic propensities, Popper may well be overlooking the way in which sub-atomic particles are, in

[12] Cf. Popper's *Quantum Theory or the Schism in Physics* (Hutchinson, London, 1982), pp. 151–6.

contrast to pin-balls, never fully isolable from the larger systems in which they operate, a fact which the talk of complementarity (which Popper dislikes) does try to do justice to.

Even if, despite Popper, quantum theory provides us with no conclusive evidence in favour of the propensity theory, that there is a genuine difference between the propensity theory and the frequency theory has been claimed on the following grounds. An experiment is performed of tossing a coin 2000 times. Prior to the experiment it is postulated that the coin has a propensity of 1 in 2 of heads appearing. From this hypothesis and a number of standard assumptions, it is possible to make predictions about the relative frequencies of heads appearing, both in the whole sequence and in certain sub-sequences. For example, the relative frequency of heads should be 0.5 ± 0.025 in the whole sequence, 0.5 ± 0.049 in the sub-sequence consisting of every fourth toss, and 0.5 ± 0.047 in the sub-sequence of those tosses which follow two tails in succession. The fact that these and other compli- cated predictions regarding observed frequencies can be made on the basis of a propensity ascription, and subjected to empirical test, shows that the propensity ascription does amount to more than simply noting frequencies.[13]

To this defence of the propensity theory, the frequency theorist would, I think, say the following. The example shows that ascribing a propensity to some repeatable experi- ment is not the same as merely *noting* frequencies. But, then neither is hypothesizing and predicting that particular gener- ating conditions will yield certain frequencies the same as noting actual frequencies. The difference between frequency and propensity theories is over the meaning of probability statements. In speaking of the coin having a propensity of 1 in 2 to land heads, are we doing more than (in effect) predicting that it will manifest various frequencies of falling

---

[13] I owe this point and the example to Donald Gillies.

heads, such as those mentioned in the example? And it is
hard to see, how, in this example, talk of propensities helps
with the single case. Let us assume that the 572nd throw
belongs to all three sequences mentioned (i.e. it is part of the
whole sequence, a fourth toss, and a toss following two tails).
The propensity theory seems to be placing it in three
different classes, each of which have different frequencies of
falling heads. Does this mean that really the 572nd throw has
three different propensities of falling heads? If it is said at
this point by the propensity theorist that neither propensities
nor probabilities should be predicated of single cases, it is
once again hard to see how the propensity theory really
differs from the frequency theory. Both seem to think of
probabilities in terms of repeatable events, and of frequencies within real or imaginary sequences of such events. The
main difference between the two theories now seems to be on
the stress given to generating conditions in the selection and
description of the sequences within which we predict and
observe frequencies. But there seems no reason in principle
why the frequency theorist should not, *qua* physical scientist,
concentrate on what can be seen as generating conditions in
selecting his sequences.

When we come to the single case, it is true that nothing
follows from the frequency theory about its outcome. But, it
is now clear, nothing follows from the propensity theory
either. And in the case of events like coin-tossing, which we
like to regard as actually determined, talk of probabilities and
propensities can well look like an oblique way of admitting
our ignorance of the determining factors in the particular
case. While we know something in general of the relevant
factors, this knowledge is only applicable to runs of events,
and we cannot weigh up the impact of the factors in any
individual case, at least not without having precise knowledge of other details of the individual case. Where the event
in question is determined, or regarded as being determined,
there is clearly something right in the subjectivist claim that

seeing it in probabilistic terms is indeed tantamount to a profession of ignorance regarding the details of the case.

The situation is quite different with a genuinely indeterministic case. If there really are no factors determining the path of a sub-atomic particle, then our not being able to predict its path is not due to any surmountable ignorance. If this is so, then a subjectivist attitude to the statements assigning probabilities to the various paths it might take would be at best misleading, for it implies that there might be something more to be known prior to the event which could enable an observer to make a reasoned prediction of its outcome. But if nature itself is probabilistic here, we would be wrong to regard our use of probabilistic theories in this area as merely a symptom of some cognitive deficiency on our part. If in nature at the sub-atomic level there are simply statistical regularities among populations of particles, whether we see these regularities fundamentally in terms of frequencies or of propensities which are held to generate those frequencies, the single event would at a deep level be unpredictable and inexplicable. Neither the frequency theory nor, we must now acknowledge, the propensity theory can deal with the single case, except as a member of some actual or virtual sequence of events, with all the uncertainty that entails over the selection of the most appropriate sequence in which to place the event. For reasons already given, it would be quite misleading to see the propensities relevant to probability theory as quasi-forces generating or determining events individually. They are actually propensities for types of repeatable set-up to manifest statistical frequencies over runs of events. If, as it surely should be, the propensity theory is understood in this way, there seems at most a difference of emphasis between it and the frequency theory of probability.

While the propensity theory stresses the generating conditions underlying observed frequencies, the frequency theory remains more epistemological, as it were, emphasizing that our only evidence for talk of propensities is observed long-run frequency and suggesting that this talk comes to no more than a reference to actual or theoretical long-run frequencies. I think it is fair to say that the arguments of the propensity theorist against this analysis are not conclusive. Indeed, to an extent, the difference between the frequency theorist and the propensity theorist is analogous to the ones we have already encountered in connection with Humeans and anti-Humeans on cause, or between positivists and realists more generally. As such, it is an aspect of a rather wider dispute which cannot be finally decided by consideration of probability alone. As far as dealing with and assessing probability statements goes, the more significant divergence is not that between frequency theorists and propensity theorists, who both analyse probabilities in terms of objective tendencies in the real world. It is rather that between objectivists about probability, which include both frequency and propensity theorists, and subjectivists, who see probabilities in terms of what we as observers are entitled to believe on given evidence, and who analyse talk of probabilities as being founded in human ignorance.

What actually emerges from our survey of philosophical interpretations of probability is that while there is a sense in which some probability statements can be regarded subjectively, in terms of our ignorance of determining conditions, this is not the case in areas in which there is genuine indeterminism. Subjectivist approaches to probability have a certain plausibility when we come to deal with the single case in the typical gambling situation so much discussed by classical theorists of probability. Though, as the paradox of ideal evidence shows, even here we would not be wrong to see the frequencies engendered by dice and coins as physically real, and to regard statements about such frequencies as not just confessions of ignorance. When we come to cases of real indeterminism, though, there is no necessary connection between the use of a probability statement and human ignorance. In so far as it seems ineradicably indeterministic, quantum theory most naturally pushes us in the direction of an objective interpretation of probability, to some version of either frequency or propensity theory. The most natural interpretation of quantum theory and its probabilistic theories is that what we are dealing with are set-ups which manifest statistical regularities, and that these regularities are both real and objective, without necessarily being based in any unknown factors determining single cases. Looked at in this way, neither quantum theory nor its associated probability statements have to be analysed subjectively in terms of the knowledge (or ignorance) of the observer.

Indeed, for most standard and scientific uses of probability statements, including those in quantum mechanics, the natural interpretation is to see the statements as referring to real frequencies or propensities in populations of particles, molecules, genes, coins, dice, and so on. On the other hand, there are also cases where we speak of probability in a more subjective way, to refer to the probability of some outcome on some evidence we have, for example, to the probability of its raining tomorrow, and in such cases it is plausible to link talk of probability to ignorance of determining conditions. Something similar would also apply to the way Bayesians speak of the probability of theories after testing, where again what is at issue is the degree of confidence some incomplete evidence entitles us to have in a theory. If this line of thought about two senses of probability is correct, then, we will have to examine specific cases to see whether the notion of probability is being used in an objective or subjective sense and, hence, whether an objective or subjective interpretation of probability is appropriate for the particular case.

8
Scientific Reductions

Scientific explanations are inherently general and, in that way, inherently reductive. A welter of diverse phenomena are brought into some sort of order by being seen as examples of some causal or statistical regularity. The differences between the phenomena are discounted or reduced, at least for the purposes of the explanation. The fall of an apple, the motion of the tides, and the orbits of the planets are all seen as manifestations of a single force, obedient to the same formulae and general principles. The source of the scientific enterprise is, as Wittgenstein said, a craving for generality. In generalizing about things, we reduce differences. In science we tend particularly to reduce manifest differences, differences of appearance. It is by this means that science is able to produce theories of wide scope and application, and so to extend our power over the world.

The reductive, generalizing tendency of science does not simply push us in the direction of theories of wider scope and universality. It also leads to what might naturally be thought of as theories of greater depth. Here we might take an example from chemistry, which leads us to think of everything in the world as being composed of a certain fixed number of basic elements in various combinations and quantities. (The number may be given as 92 or 103, or some other number, depending on the definition of element, and the current state of chemistry.) And then physics teaches us that these elements themselves are constructed from a smaller number of basic particles, again in various combinations and permutations. We thus approach the age-old dream of seeing everything in the universe as constructed out of some elemental building stuff, from fire or from water or from love and strife, or, more scientifically, from atoms moving in the void, according to determinate and discoverable laws.

It will be noted that the direction of a reduction is characteristically downwards, penetrating beneath appearances to an ultimate material out of which all else is composed and in terms of which the behaviour and appearance of all else is to be explained. This is partly why reductionist programmes can be seen as converging ultimately on physics, for physics is taken to be the science which deals with the ultimate make-up of things, their most basic constitution. But physics is also the science which addresses itself to the constitution of all matter. Even if we reject the idea that the biological, say, can be reduced to the physical, and believe that life represents a step beyond what is wholly explicable in terms of the laws of physics, it is still the case that biological organisms are made up of particles of matter, and to that extent are subject to normal physical processes, as we see when a living body falls or is exposed to radiation. On the other hand, non-living things are not subject to biological processes, except accidentally, as when, say a man or an animal moves a stone in the course of achieving some biologically specific goal. The scope of physics, then, is unlimited, while that of biology and the other life sciences (including the social sciences) is restricted to rather special domains. If a reduction of biology and the life sciences to physics were achieved, then what we would be saying in effect would be first that life, consciousness, and society are explicable simply in terms of the behaviour of their basic material constituents and require no special sort of explanation invoking the emergence of special types of property in special circumstances, and secondly that they are thus in essence no different from everything else in the universe. Even though the biological realm looks different from the inorganic, this is simply a difference of initial appearance which conceals an underlying and essential unity of process and nature. Such a conclusion would surely be of some significance, not least in terms of how we might think of ourselves as human beings.

In science, then, there is and has been a quest for theories of wide scope and profound depth, the hope being that the possession of such theories will yield a unified account of the construction of things in the physical world, and of their operation. Nevertheless, in the history of science the hope has actually been more of a dream, inspiring research, no doubt, and leading to all sorts of discoveries and new knowledge, no doubt, but not to the complete fulfilment of the hope. Attempted reductions have characteristically broken down when seemingly close to their goal, and led to subsequent shifts of paradigm or metaphysical world-view.

Reductions in the Physical Sciences

Karl Popper has written eloquently about the history of physics and chemistry in terms of attempts at reductions.[1] Before considering criteria for reductions more formally and touching on some of the controversial issues involved in attempts at the reductions in the life sciences and the human sciences, we will look at some of Popper's examples of reductions in the physical sciences.

Popper's basic thesis is that programmes of reduction are part of the 'activities of scientific and mathematical explanation, simplification and understanding',[2] and that a reductionist programme can fail without being a failure. A reductionist programme fails if it fails in some way to show that the reduction envisaged can go through. A familiar

[1] Particularly in *The Open Universe* (Hutchinson, London, 1982), pp. 131–74.
[2] *The Open Universe*, p. 134.

example here, from mathematics rather than from the physical sciences, would be the logicist programme associated with Russell and Frege: the attempt to show that arithmetic could be reduced to logic, together with the intuitive notion of a class. This programme is generally regarded to have failed both because of the paradoxical implications of the intuitive notion of a class or set, and because of Gödel's proof of the incompleteness of arithmetic. The former difficulty arises because there are well-defined predicates (such as the predicate 'is not a member of oneself') which lead to paradoxical classes (the class of all classes which are not members of themselves). Attempts to modify the notion of a set to take account of this problem appear to take us beyond what could reasonably be regarded as falling within the realm of logic. Gödel's proof implies that mathematics as a whole cannot be codified in a single formal system and that even that part which can be expressed in a specific notation cannot be so codified. Any formal arithmetical system will contain statements which are true but not provable in the system. Gödel's result undermines the logicist hope that mathematical truth is a property which true formulae in a system derive deductively from the premisses of the system.

Logicism then demonstrably failed. Arithmetic could not rest on such slender foundations. But despite the failure of logicism as a reduction, one might quite reasonably contend that a lot had been learnt through the attempt and the discovery of its impossibility. In this sense, the programme failed without being, intellectually or cognitively, a failure. And this is how Popper sees many of the most famous attempted reductions in the history of the physical sciences.

Indeed, in his writings, Popper contends that there have been hardly any completely successful major reductions in science. The two he draws attention to as being almost completely successful are, in fact, not actually typical of reductions in the physical sciences. They are what Popper calls Newton's reduction of Kepler's and Galileo's laws, and the reduction of rational fractions to ordered pairs of natural numbers. The latter is, of course, an example from mathematics rather than from the physical sciences, and even this fell short of the original intention of the ancient Pythagoreans who had first embarked on a programme of showing that natural numbers underlay all natural phenomena. The discovery that the square root of 2 cannot be expressed as a ratio of two natural numbers was fatal to the Pythagorean programme, and yet such a number would be the length of the diagonal of a square whose sides are one unit in length. So we have a real physical magnitude (the length of the diagonal of the square) which is not, in Pythagorean terms, arithmetically expressible. How then could number be the essence of the world?

What Popper calls the Newtonian reduction of the laws of Galileo and Kepler is a considerable scientific achievement. Newton's laws unify two previously disparate areas of knowledge, Galileo's terrestrial and Kepler's celestial physics. In effect, Newton is saying that the earthly projectiles of Galilean theory and the planetary motions of Kepler are examples in action of the same laws of dynamics. The same forces, in other words, are in operation in both cases. Now while Newton's theory has greater universality than either Kepler's or Galileo's, it is not clear that what is at issue here is a reduction in the full sense. The forces postulated by Kepler and Galileo are not being broken down into simpler components, despite the obviously greater scope of Newtonian dynamics. So this case is not a straightforward or typical scientific reduction, unlike, say, Cartesianism or atomism.

Popper analyses the Cartesian reduction in the following terms. In it, the whole of the physics of inanimate matter was seen in terms of extended substance:

the Cartesian physical universe was a moving clockwork of vortices in which each 'body' or 'part of matter' pushed its neighbouring part along, and was pushed along by its neighbour on the other side. Matter alone was to be found in the physical world, and all space was filled by it. In fact, space too was reduced to matter, since there was no empty space but only the essential spatial extension of matter. And there was only one purely physical mode of causation: *all causation was push*, or action by contact.[3]

This picture is clearly simple to understand and intuitively compelling, so much so that Newton himself at one time felt compelled to explain gravitational attraction in its terms, as the impulse of cosmic particle bombardment. This way of saving the Cartesian reduction, though, conflicted with Newton's own basic principle of inertia, because it would mean that all moving bodies would be operating in a braking medium of rain-like particles impeding their movement. So the Newtonian postulation of a fundamental mode of action at a distance eventually proved fatal to the Cartesian reduction, and those more Newtonian than Newton himself began to regard gravitational attraction itself as a fundamental property of matter.

Newtonian theory naturally had its own reductionist impetus, but by the turn of our century it had fallen foul of the inability of physicists to see electricity and magnetism in terms of Newtonian forces. Instead, for a number of years, the opposite programme held sway. Popper quotes a passage from Einstein's *Sidelights on Relativity* of 1922, which states that 'according to our present conceptions' the elementary particles (i.e. electrons and protons) are '*nothing else* than condensations of the electro-magnetic field', and that 'our' picture of the universe presents 'two realities', gravitational ether and the electromagnetic field, or 'as they might also be called—space and matter'.[4] The discovery of nuclear forces, of course, put a stop to this particular reduction, but, as Popper points out, until 1932, when the neutron was discovered, nearly all the leading physicists subscribed to the reduction of all matter to two elementary particles, the proton and the electron.

[3] *The Open Universe*, p. 135.
[4] In *The Open Universe*, p. 139, from *Sidelights on Relativity*, p. 24.

The analysis of everything in the universe in terms of gravity and electromagnetism, and the reduction of all matter in the universe to electromagnetic forces condensing into two elementary particles has been shattered not only by the discovery of further elementary particles, such as the neutron and the positron which had been identified by 1935, but even more by the discovery of apparently irreducible nuclear forces. Whereas it had been hoped to explain gravity in terms of general relativity, and to bring both gravitational and electromagnetic forces under a unified field theory, we are now faced with a plethora of elementary particles, and with four basic forces (gravitation, weak decay interaction, electromagnetism, and the strong nuclear forces). Even now, impelled by reductionist dreams, physicists are working on new unified field theories, but we must wait to see the extent of their success and comprehensiveness.

Whatever might ultimately be the case in physics itself, it is certainly widely believed that there has been a successful reduction of chemistry to quantum physics. The claim is that we can account in terms drawn from quantum physics for the make-up and behaviour of the various chemical elements. Bohr's quantum theory of the periodic system of the elements not only predicted the chemical properties of elements, but also the existence and properties of an as yet undiscovered element (to be called Hafnium, when it was found in 1922). Popper points out that the discovery of heavy water entailed that the measurements of atomic weights, and hence Bohr's tables, were wrong. Bohr's tables were reformulated in the light of this, but even so it is not clear that quantum theory can actually explain the nature of the chemical bond or certain features of the periodic system of the elements. Let us though assume that these and other problems could be overcome within quantum theory: would we then have a satisfactory reduction of chemistry to physics?

Popper argues that in a very significant sense we would

not. The reason for this is that the heavier nuclei are composed of lighter ones, down to the lightest of all, that of hydrogen, which is regarded as the basic building block. The heavier nuclei (i.e. those of all other elements) are formed by fusion of hydrogen nuclei. But this process, starting with the transformation of heavy hydrogen into helium, happens only in very rare conditions of great temperature and pressure in which the mutually repulsive electrical forces of the nuclei are overcome by the nuclear force. These conditions generally obtain only in supernovae explosions, but helium forms about one-quarter of the matter of the universe. Hydrogen forms between two-thirds and three-quarters, and so the rest of the heavier elements are very rare altogether, perhaps only 1 or 2 per cent by mass.

So how did so much helium come about? At this point, according to Popper, physical theory has to be supplemented by cosmology and cosmogony. The hot big bang of the first minutes of the universe is regarded as the source of most of the helium. But invoking a unique and uncertain event in order to build the bridge between physical theory and the existence of the chemical elements is, in Popper's view, hardly a convincing basis for a claim that a reduction has been fully executed. Moreover, the nuclear bonding property of hydrogen nuclei is inoperative in most of the conditions which exist in the universe, and could hardly be postulated on the basis of what Popper calls the manifest physical properties of the hydrogen nucleus. Furthermore, in postulating the creation of helium and the other elements, we are asserting that in certain very rare conditions the gravitational force, which is weaker than and apparently unconnected with the nuclear and electrical forces, can overcome the electrical repulsion between the nuclei in such a way as to bring them into a position to fuse by virtue of the nuclear force. Once again, a potential crucial to the creation of matter other than hydrogen—in this case the harmonizing of two unconnected forces in very rare conditions—is something which could hardly have been predicted on the basis of the manifest physical characteristics of the various forces and hydrogen.

Popper's conclusion is that the reduction of chemistry to physical theory assumes a theory of the origins of the universe in order to allow 'sleeping potentialities, or relative propensities of low probability built into the hydrogen atom to become activated'. In fact, he thinks that we would do well to recognize that in speaking of the fusion of the hydrogen nucleus and all that is entailed by it, we are really speaking of what he calls an emergent property, unforeseeable outside the very special conditions in which it obtains.[5] Popper's view is that the reduction of chemistry to physical theory is a reduction to a physics which assumes cosmology, cosmogony, and the appearance of emergent properties in special and unusual conditions, in order to explain the actual existence of the chemical elements in their quantity and range. As such he would describe it as a reduction which is far from complete. Presumably, Popper is not claiming that in the special conditions involved the laws of physics are contravened, but only that in such conditions states of affairs come about which would appear so improbable on purely physical grounds that we would be justified in speaking of emergent properties here. Whether he is right on the relation of chemistry to physics and on the nature of emergent properties—and his reductionist opponent will presumably deny that rarely manifested effects cease to be manifest physical properties and become emergent chemical properties, just because they are manifested only in rare and special conditions—Popper has certainly raised a number of issues which will become significant in considering other and even more controversial candidates for reduction. It would therefore be as well to look more systematically at just what is involved in a reduction of one area of science to another.

[5] Cf. *The Open Universe*, p. 145.

Criteria for Reduction

A scientific reduction will characteristically operate between two different levels of theory or discourse. The aim of a reduction is always to show that one level can be explained in terms of the other. The Greek atomists, for example, wanted to show that everything that existed was ultimately merely the operation of atoms in the void. Chemistry, we have seen, is widely supposed to be reducible to (or actually reduced to) physics. That is, chemical reactions are held to be particular cases of underlying physical processes, explicable in terms of these processes, and chemical substances to be composed of the types of particles mentioned in physical theory in accordance with physical laws and processes. And, of course, the reducibility of biology to physics (and chemistry) and of psychology and sociology to the physical sciences are controversial and much discussed issues.

One point that emerges immediately from the previous paragraph is that the question of a reduction in a given area cannot be assessed without reference to our schemes of explanation and classification. There is no unproblematic pre-theoretical sense in which some things are definitely chemical things and others clearly physical, and this is part of the problem we have in assessing Popper's claim about the non-reducibility of chemistry to physics. His opponent would obviously reply that he is making too much of the idea of very special conditions and unlikely coincidences, and would claim instead that potential for fusion in hydrogen atoms is as much a physical force as any other, and not an emergent 'chemical' property. Popper's opponent would hold that while we might, for convenience and ease of explanation and manipulation of specific chemical processes, continue to use chemistry and its laws, we would do this while recognizing that these chemical processes were really governed by the more fundamental laws of physics and explicable in their terms.

In a way, indeed, the whole point of a reduction is to deny that classifications demarcating chemistry and physics, say, or biology and chemistry, have any real grounding in reality, and to blur the distinctions between the different levels. But even if one takes an anti-reductionist stance on some issue, there is no secure pre-conceptual demarcation between one level and another. To put this point concretely, we have just spoken of the attempt to reduce chemistry to physics. But what does this mean? What would such a reduction actually consist in? The answer must be in terms of the *disciplines* of chemistry and physics. The reduction will show that the things spoken about in chemistry can, in some sense to be determined, be explained in terms and laws drawn from physics. And here a further complication arises, for the disciplines of chemistry, physics, and the rest are not timelessly given, nor are their boundaries static over time. No doubt people have an intuitive sense that physics and chemistry deal with non-living matter and biology with living matter. But is it clear that, if some quantum theorists and parapsychologists have their way, life and even consciousness could never be seen as, vestigially at least, a property of all matter, and hence admissible in an unreduced way into some future 'physics'? Outlandish as this speculation might seem, I introduce it simply to show that even the boundaries of physics—the *terminus ultimus* of the reductive enterprise—are not fixed. After all, at one time, the notion of action at a distance was not counted as a genuinely physicalistic property, and was regarded as a subject more suitable for magic or theology.

So there is going to be a degree of relativity to changing human constructions and distinctions involved in any talk of a reduction. This, though, does not prevent us from specifying some fairly clear criteria for reductions. Taking a given area of knowledge as it is at a certain time, we can speak of its having been reduced to another level of enquiry in at least three distinct respects.

The first of these respects will be when the terms of a given level of enquiry can be shown to be extensionally equivalent to terms from some other level of enquiry. By this I mean to rule out any suggestion that a scientific reduction of terms implies that a term from biology, for example, which has been reduced to some chemical term, has the same meaning as the chemical term. To say that genes are DNA molecules is not to say that 'gene' means 'DNA molecule'. Quite clearly, finding that genes were DNA molecules was a matter of empirical discovery, and a highly impressive one at that. Nevertheless, 'gene' has a particular role to play in a specific explanatory framework (= unit of replication), a role and a connotation which does not attend the term 'DNA molecule', which is a way of referring to the molecular make-up of the gene, and not to its functional role in the transmission of hereditary characteristics. However, differences of meaning notwithstanding, the discovery of the extensional equivalence of genes and DNA molecules is an example of successful reduction of terms from biology to chemistry, and indeed a major element in the confidence many have that biology as a whole might be reducible to physics and chemistry.

Reducing the terms of one level of explanation to that of another is not, however, sufficient for a successful scientific reduction. Even if all biological phenomena, let us say, have physico-chemical substrata and so physico-chemical definitions can be given for biological terms, it may still be that their behaviour in biological set-ups has properties and characteristics which could not be predicted on the basis of the laws of physics and chemistry. New or emergent properties of already existing types of matter might come into play in certain special circumstances; new types of behaviour may be manifested by the original stuff when operating in special conditions. This is the point Popper is making about the fusing of hydrogen atoms to produce the chemical elements, and no doubt analogous points could be made about the behaviour of physico-chemical material within a living organism. For this reason, it has seemed to many that a successful scientific reduction will require the reduction of the laws of the higher level to those of the lower level, as well as the reduction of terms.

Following Hempel,[6] we can analyse the reduction of laws from one level to another in the following way. Let us suppose we have a biological law which says that whenever we have B_1 we have B_2, where 'B_1' and 'B_2' are terms referring to some entity or property described in biological terms. If we then had some suitable reductions of the biological terms to physical terms, 'B_1' to 'P_1' and 'B_2' to 'P_2', and a physical law saying whenever we have P_1 we have P_2, we would clearly have a reduction of laws of the biological level to the physical level. Indeed, the biological law would, in these conditions, be logically deducible from the physical law. Of course, the reduction depends first on the existence of a reduction of the relevant terms (in effect, connecting laws saying that whenever we have B_1 we have P_1 and whenever we have B_2 we have P_2), and, secondly and crucially, on the existence of the appropriate physical law. Opponents of postulated reductions will often focus on the absence of such a law. They will say that there is no principle derivable from the principles of physics alone which links occurrences of P_1 to occurrences of P_2. This, in effect, is what Popper is doing in his criticisms of the proposed reduction of chemistry to physics. Proponents of the reduction, on the other hand, will assert that the fusing potential of hydrogen atoms in conditions of extreme pressure and temperature is a purely physical property, and quite properly to be included in physics.

It will be obvious that whether we ever have a scientific reduction in the sense just explained depends on empirical discoveries, on finding reductions of the relevant terms and laws, and this is clearly going to be an onerous undertaking

[6] In *The Philosophy of Natural Science* (Prentice Hall, Englewood Cliffs, NJ, 1966), p. 105.

where we have two well-developed levels of description, even assuming we allow a certain tolerance in mismatches of either terms or laws at the edges. Now, while no one would doubt that we would have good grounds for speaking of a reduction if the conditions are fulfilled, it might be wondered whether we might sometimes be justified in speaking of a reduction in cases where the conditions are not fulfilled. In other words, are we entitled to talk of reductions only where we have reductions of terms and laws as envisaged by the two conditions just considered?

As an example of how laws at one level might not be reducible to laws at another level, we might consider the well-known generalization that you cannot get a square peg one inch square into a round hole one inch in diameter.[7] In so far as we have a proof of this generalization, it will be based on geometrical principles, as indeed is our understanding of the terms 'round' and 'square'. It is, to say the least, unlikely that we could get any reduction of the geometrical notions of round hole one inch in diameter or square peg one inch square to physicalistic terms. After all, the physical realizations and underlying molecular structures of square pegs and boards with round holes are going to differ immensely, depending on the materials involved (wood, plastic, metal, etc.). And it is even less likely that, assuming we found some physicalistic reduction of the terms involved, there would be any law or laws in physics which did not refer to the geometrical features of the case which expressed the simple truth that the one sort of thing could not go into the other. The point is that whenever you have a structure with the relevant geometrical properties and rigidity in peg and board, you get impenetrability, whatever the underlying molecular structure of the materials involved. At the same time, though, it would be rather odd to think that there was

[7] The example is derived from Hilary Putnam's 'Philosophy and our Mental Life', in his *Mind, Language and Reality* (Cambridge University Press, 1979), pp. 295–7.

anything in an instance of a square peg failing to go into a round hole which was not physical through and through and explicable in physicalistic terms—such and such a lattice of particles preventing the free movement of some other lattice. It is just that we cannot express our geometrical generalization covering all instances of square pegs and round holes of the relevant sizes merely by referring to underlying molecular structures, although we could undoubtedly express physicalistically the impermeability of any particular square peg and round hole.

The non-expressibility of the generalization about square pegs and round holes in physicalistic terms would mean on our original two conditions for a reduction that we had failed to reduce the statement to physics. Yet, no one would want to say that this shows that there is something about square pegs failing to go through round holes which transcends physics. Those who say such things about consciousness or life or the creation of helium would certainly not want what they say to be regarded as analogous to the example of round holes and square pegs. In the former cases, but not the latter, there is a question of some new level of reality, of an emergent property at least, of a hitherto unpredictable sort of behaviour. The difference between the two types of case is surely that in the square peg and round hole case no one doubts for one moment that a physicalistic explanation could be given for every individual case falling under the geometrical generalization. Only a reduction of the generalization itself cannot be given without resource to geometry.

We should, then, weaken our original conditions for reduction to allow for cases like the square peg and round hole. What we might now say is that a reduction takes place *either* if terms and laws are reduced *or* if every state of affairs covered by the level of explanation to be reduced is also describable and explicable in terms and laws drawn from the reducing level of explanation, without insisting on any isomorphism of laws or connecting principles linking terms

from one level of explanation to the other. Reduction understood in this sense would be a reduction not of one level of
explanation to another but a reduction which showed that
some phenomena could be explained in different and irreducible ways. The geometrical explanation stands unreduced, and in this form will serve all sorts of purposes.
Indeed, it would usefully link all sorts of cases of square pegs
and round holes which no physicalistic account would group
together. But the irreducibility of the geometrical level of
discourse does not imply anything about levels of reality not
governed by the laws of physics: we will continue to suppose
that all actual square pegs and round holes are so governed.
In this sense we keep a physicalistic ontology, while admitting the existence and usefulness of irreducible levels of
description and explanation of some physical phenomena.

It would surely be right to insist on this third and wider
sense of reduction in order to assess reductionist claims. For
these claims are fundamentally about levels of reality and the
existence or non-existence of emergent properties or entities,
of 'new' sorts of things or behaviour. The peg and hole
example shows that we may well have and use irreducible
levels of explanation and description without any commitment to any corresponding emergent properties or existences
of a new sort. An extended sense of reduction becomes
extremely relevant once it is realized that the types of
explanation offered at one level of description may not be
susceptible of any straightforward reduction to the explanatory framework of other levels.

A now familiar though rather controversial example in this
context would be the claim that psychological explanations
involving reference to mentalistic activities and states are
essentially functional and refer to systems of behaviour
rather than to the inner structure of persons or brains or
other bearers of psychological predicates. Thus, it might be
argued that adding is an activity performed indifferently by
human beings, electronic digital computers, and mechanical

calculators. Whether something can correctly be described as adding depends simply on its yielding certain outputs when given certain inputs. There is no implication that different types of adding agents will have the same physical make-up or that different cases of adding have the same physical realization. Even in the case of different human beings adding the same sums, it would be most implausible to suppose that they were in the same brain states, given their very different life histories and genetic endowments. It is then most unlikely that there will be any single physicalistic predicate coextensive with or even underpinning all cases of adding, or that any conjunction of such predicates could be brought into any law-like combination. In fact it might be claimed that there is a sense in which the functional description and the physicalistic explanation are complementary, the latter describing the mechanisms underlying specific instantiations of the functionally described activities. And the same goes for terms referring to other activities and psychological states. In the case of the computers and calculators, there is no suggestion that there is anything in their make-up not explicable in physicalistic terms. Opponents of physicalistic reduction of the mental might wish to claim at this point that it is wrong to describe machines as adding. They 'add' only in virtue of their being programmed to do so by human beings who can intend to add and to programme machines to do so, and this human ability to add and perform other mentalistic tasks cannot be explained in purely physical terms.

The example of mentalistic predicates, like that of square pegs and round holes, shows us that even if physicalism were true, and all phenomena can under *some* description be explained by the laws of physics, there is still a point in using levels of description and explanation of a non-physicalist sort. Through their use we can come to see similarities of structure and behaviour where descriptions drawn from physics could discern only differences at the material level.

Nevertheless, as already remarked, having levels of explanation irreducible to and, in a way, autonomous of physics is not enough to establish that anything which happens in the world cannot under some description be shown to be happening in accordance with the laws of physics. Simply by pointing to a different level of description and explanation, we have not shown that objects start to behave in ways or manifest properties which could not under some description be captured and predicted by the laws of physics. That the use of mentalistic predicates to describe the behaviour of persons does not imply that at some level every movement of and within the organisms we are calling persons is not predictable and explicable by means of the laws of physics is, of course, just what physicalists about persons will want to insist on. And from what has been said so far, they are right to insist on it, because irreducible levels of description on their own are not enough to rule out genuine reductions, especially if, as would be claimed here and as is the case with square pegs and round holes, differences of state at the higher level of explanation 'supervene' or depend on differences at the lower level (though without any laws linking types of mental state with types of physical state).

So far, mental states have been seen as primarily functional states; while this may be plausible in the case of physically manifested mental states, such as adding or believing that it is going to rain, there are, of course, also the subjective experiences of conscious beings, their actual pains, thoughts, and the rest. A philosopher of a generally materialist-reductionist disposition might not seek to deny the existence of subjective experiences, but to treat them rather as mere epiphenomena, depending on physical states for their existence but not at a fundamental level causally effective in themselves. We will consider this type of approach further in the next chapter.

What is fundamentally at issue in considering scientific reductions is whether there are in nature different levels of existence and complexity, with a higher level being irreducible to a lower one. Thus we can ask whether chemistry is reducible to physics, whether biology is reducible to physics and chemistry, whether psychology is reducible to biology and physics and chemistry, whether sociology is reducible to psychology and the rest, and perhaps also whether human behaviour is susceptible of a complete scientific explanation of any sort. It is obviously important in any of these cases to see whether reductions of terms and laws can be carried out, but irreducibility here does not mean we have to assert that, say, the properties of biological processes are emergent over physical and chemical ones. The irreducibility of biological explanations to those of physics and chemistry does not on its own mean that things in the living world behave in ways inexplicable in terms of their physics and chemistry.

In fact, though, there may be reasons within biology to suppose that certain types of causality operate only in conditions of a certain complexity, and could not be foreseen by studying situations of less complexity. At least, there is good evidence that biological wholes are more than the sum of their parts. In natural selection, some causal mechanisms operate only on a level involving groups of individuals. There is, for example, the phenomenon of polygeny, whereby single genes have specific effects only in combinations with other genes. Depending on which other genes a gene is joined to, the probability of a given effect can be either raised or lowered. The consequence of this is, in the words of Elliott Sober, that 'there is no such thing as the causal role that the gene has in general'.[8] In other words, ensembles of genes are causally efficacious where individual genes are not.

Similarly, it is argued that some of the mechanisms involved in species selection are distinct from and not reducible to individual selection. The point here is that there can be characteristics operative in the survival and distribution of groups of creatures which have no bearing on the

[8] *The Nature of Selection* (MIT Press, Cambridge, Mass., 1985), p. 313.

survival of individual members of that species. In this context, Sober cites Stanley's claim to explain by reference to group selection why there are a large number of small species of wingless grasshoppers and only a small number of large species of winged grasshoppers.[9] While having wings or not apparently has no effect on the potential for the survival of individual grasshoppers, if we envisage an original population of grasshoppers consisting of both winged and wingless members, we can see that the wingless ones will have less tendency to roam and mix and so tend to form small, isolated sub-groups which will speciate. Once again the claim is that a causal mechanism comes into effect only at the level of the group, and cannot be identified with anything at the individual level.

If within biology there are group effects not reducible to the individual effects of individuals within larger populations considered in isolation from those larger populations, it may be no less plausible to think of life itself as a process or mechanism emerging out of physical and chemical elements in very special conditions and combinations. The anti-reductionist will not necessarily be committed to a vitalist position, that living things are composed of something other than chemical and physical stuff, but he is committed at least to the position that, in living organisms, chemical and physical stuff behaves in ways unforeseeable on physico-chemical grounds, and that what we have here is an unforeseeable evolutionary development (rather as Popper claimed that the 'cooking' of the heavier elements was physically unforeseeable and, hence, emergent). The claim that there are biologically emergent properties is that, when in specific circumstances and combinations, certain types of physico-chemical material produce behaviour or effects which could not have been foreseen from its behaviour outside the new and special domain represented by our talk of the biological.

---

[9] Sober, *The Nature of Selection*, pp. 366-8.

Thus we would not be wrong to see an analogy between the way non-reductionists see new causes and types of behaviour at work, once the line between the physico-chemical and the biological is crossed, and the way in which, within biology, wholes, such as groups of genes or groups of grasshoppers, introduce causal mechanisms not attributable to their individual components acting on their own. This analogy would be rather agreeable to a cautious anti-reductionist anxious to avoid vitalist conclusions that living things do not only have non-physico-chemical properties, but that they actually contain some sort of non-physico-chemical stuff. The ensemble of genes and the group of grasshoppers are undoubtedly made up of genes and grasshoppers, and nothing else. In the same way, we might envisage living things being composed of physico-chemical stuff, and nothing else (and so reject the vitalist thesis of some special type of living stuff), while seeing the physico-chemical stuff of which living things are made as engaging in new or emergent types of behaviour.

In speaking of new or emergent types of behaviour, we would be thinking of the behaviour of biological organisms, or of groups of genes or grasshoppers, as unlike that of physical examples of square pegs and round holes. In the latter type of case, there is no suggestion of emergence of new properties or behaviour owing to the fact that the pegs and holes are in conditions of squareness and roundness. But, as we have seen, no one would argue there for anti-reductionism.

Any anti-reductionist claim will always involve the idea of new types of behaviour at least. Some stronger types of anti-reductionism will involve the postulation of new types of stuff, such as life-stuff or soul-stuff, to explain newly emerged properties or behaviour. However, our topic is *scientific* reductions, and life spirits or souls may well evade the scientific net. Scientifically, the first and most significant step will always be the examination of the claim that in certain specific conditions we find material objects behaving in ways unforeseeable on the basis of the behaviour of material objects outside those very special conditions. Scientific rather than philosophical argument in this area will centre on the adequacy of existing or reasonably foreseeable scientific theories in explaining the behaviour in question.

It remains, though, a moot point as to just how we should conceive emergent behaviour, if we should find such a thing. Are we to regard it as a contravention of the laws of physics, in the way that the existence and persistence of complex living things might initially appear a contravention of the second law of thermodynamics, according to which all material objects tend to become more disordered? Or should we seek to show that with a more complex understanding of the relevant physical law, there is no contravention, as when it is said that the second law predicts growth of disorder only in closed physical systems and does not rule out the existence of open systems which export their potential for disorder (or entropy) into the environment?

Strictly speaking, of course, there can be no exceptions to the laws of physics. Exceptions would show that what we had was not really a law. And thus there is something potentially misleading in H. H. Pattee's[10] claim that molecules in living systems 'impose variable constraints on the motion of individual elements' and thus have 'an effect which is like modifying the laws of motion themselves'. While one can see what Pattee means, the thought would be better expressed by saying that if the phenomenon he is speaking of actually exists, it shows that the laws of motion are not as unrestricted in their application as we had formerly believed, and that, as in the example of open thermodynamical systems, the law has to be modified to allow for certain specific conditions in which it operates in a different way. The idea of emergence of behaviour would be preserved by insisting that, from the

---

[10] H. H. Pattee, 'The Problem of Biological Hierarchy', in C. H. Waddington (ed.), *Towards a Theoretical Biology* (3rd ser.; Edinburgh University Press, 1970), p. 127.

lower, more general level of explanation and of matter, there was something improbable and unpredictable about the new type of behaviour and the special conditions. We do not need to think of emergent properties as involving self-contradictory talk of contraventions of basic physical laws.

The reductionist will naturally resist the line of thought of the previous paragraphs. He will insist that the special effects and mechanisms of physico-chemical matter in living organisms are indeed nothing but properties of physico-chemical matter. His case would be the stronger the more he was able to demonstrate that the behaviour of living matter could, despite initial appearances to the contrary, be accounted for in terms of the same laws and processes that governed the behaviour of inorganic matter.[11] There would, in his view, be nothing new or emergent here, nothing which actually demanded explanation, beyond the same principles which could be seen to be operative at the inorganic level. The claim would be that for convenience, for special purposes, we might look at living things in terms of specifically biological laws and processes, as with looking at square pegs and round holes from a geometrical point of view, but we could equally, if more far laboriously, explain the physical movements of biological beings (including their mental states in so far as they were physically realized) by invoking nothing more than principles and laws drawn from physics and chemistry.

Analogous moves would be made over other candidates for reduction. The behaviour of individuals in society might be examined to see if it could be explained in terms of individualistic psychology. Or it might be claimed that psychological explanations of human behaviour could, as far as the movements of human bodies is concerned, be replaced by

[11] Although if the only way he was able to do this was to show the inorganic matter was in some sense alive or conscious we would be presented with a rather Pickwickian sense of reduction. Rather than biology being reduced to physics, physics itself would have been taken up into biology, in line with the parapsychological speculation referred to earlier.

neurophysiological accounts of bodily movement. In the biological realm, explanations in terms of purpose and function are often regarded as, in principle, replaceable by reference to purely mechanistic processes and talk of experiences and other mentalistic phenomena eliminable from a causal or explanatory point of view. Whether these and other programmes for reduction can actually go through is as much an empirical as a philosophical question. Much depends on the discoveries of empirical science, and on the boundaries between the different sciences, which are themselves affected by empirical discoveries. Within the scope of this book, however, it is only possible to indicate what sorts of issue might arise in considering claims that a reduction of one area to another was in the offing. What we have tried to suggest is that while a reduction by means of reductions of terms and laws would be sufficient to claim a successful reduction, such a complete form of reduction would not be necessary to satisfy the main end of a scientific reduction. Nor, given the irreducibility and lack of fit between different levels of explanation, is such a complete form of reduction likely in many cases. But the reductionist ideal would still be fulfilled if it could be shown that, under some description, the entities and processes of a higher level of discourse were in fact explicable and predictable in terms of a more basic level. And in many current discussions of reductions of the human sciences, it is this looser type of reduction which is actually envisaged.

In conclusion, it is worth stressing that were a reduction of any sort to go through, it would not necessarily imply that the laws and statements of the 'higher', reduced level of discourse were false. Indeed, a successful reduction of laws would even be a way of confirming the truth of the reduced laws by showing them to be a special case of some more general laws. But even while a reduction does not imply the falsity of the higher level of discourse, it surely does imply that there is something metaphysically otiose about that level, that its terms and laws do not describe anything not analysable and explicable in more fundamental and more general terms. Such a conclusion would imply that, while the laws and concepts of the reduced levels of discourse are not thereby falsified, any pretensions that that level of discourse had to represent an emergent, irreducible level of reality rested on an illusion.

9
Science and Culture

In the chapters of this book, we have been considering
science as an intellectual activity largely on its own terms.
We have looked at problems and questions which stem from
within the activity itself: its methods and proof procedures,
the postulation of unobservable forces and entities and its
claim to discern the deep structure of the world, the way in
which knowledge grows in science, the significance of a
scientific reduction, and the meaning of probability statements in science. All these issues are internal to the practice
of science, arising quite naturally from its theories and
procedures. But there are also questions to be asked about
science from an external point of view, about its place in our
culture as a whole, and about the relationships which may
exist between science and other aspects of human knowledge
and culture. In this chapter we shall be concerned with some
of these questions, returning to issues first touched on at the
start of the book.

Science as Mythology

August Comte argued a century and a half ago that science
could give us the 'positive' knowledge which would allow us
to displace the earlier more primitive and mythological
attempts of religion and metaphysics to provide systems of
thought for coping with experience. Comtean positive knowledge was systematic knowledge of empirical laws, based on
and checked through observation and experiment. In con-

trast to theology and metaphysics, positive knowledge was to avoid speculation beyond the observationally endorsable, including speculation of a quasi-metaphysical sort into the unobservable essence of material things. It is, of course, as a result of Comte's endeavours that the much maligned and misunderstood term 'positivism' has gained currency. Beyond remarking that many people who would reject the label 'positivist' would none the less be positivists in Comte's sense, I will say nothing more about positivism. Rather, I want to suggest that far from having displaced mythology in our world, science itself has become a mythology, perhaps the prevailing mythology of our time. While religion and metaphysics may, as Comte hoped and predicted, have fallen into disrepute in right-thinking intellectual circles, science itself has begun to take on many of the features of a mythology.

For one thing, despite Comte, modern science has not eschewed metaphysical speculation into the essence of the world. Indeed, one may be glad that science can still be seen as pursuing the ancient and noble goal of knowledge and understanding of the world for its own sake. In so doing, questions inevitably arise as to the origins and nature of the universe, of matter itself, of time and space. Much of what scientists tell us of these things inevitably goes way beyond anything we have evidence for. Any evidence we have is necessarily drawn from a tiny part of the whole universe, and may not be representative or indicative of the whole. Of course, scientists continually attempt to widen their empirical base, as we suggested in Chapter 2 when considering inductive proof. Unlike religious and metaphysical myth, there is always an attempt to subject scientific theory to empirical test. But we need to recognize the extent to which, even in science, evidence underdetermines theory, particularly when we are thinking of the grand cosmological and cosmogonic theories. None of this would matter, of course, were it not for the tendency we find in the contemporary

mind to think that science and science alone gives access to
the ultimate truth about man and the world. And yet, as we
have seen, the truth-bearing potential of science falls very far
short of anything like that. To repeat the lesson of our survey
of the philosophy of science, science does very well on local
regularities, empirical discoveries, and technological effects.
Here its claims to knowledge are strikingly well-grounded
and productive. Beyond this empirically sustainable body of
truths, though, its claims to knowledge are far more tentative
and justify far less confidence. As just remarked, this only
matters because of the tendency of scientists and non-
scientists to regard the wider claims of science as far better
grounded than they are. And the elevation of fascinating
speculation into absolute truth is one of the marks of a
mythology.

The mythology of science is perhaps particularly danger-
ous when it comes to considering questions of reduction. It is
widely asserted by philosophers at least that in the fields of
the mental and the human the doctrine of eliminative materialism is true. This is said to be required by the imperatives
of scientific realism, and involves the claim, mentioned
briefly in the last chapter, that we should regard our everyday psychological talk about people's mental states and
emotions and intentions as both dispensable and misleading.
Such talk is said to be dispensable because a fully developed
neuroscience will enable us to predict all the activity of the
bodies of persons without reference to the idioms of what is
disparagingly referred to as 'folk' psychology (our everyday
talk about persons and the mental). Folk psychology is
misleading in so far as it has connotations of dualism, free
will, and so on. The categories it uses, such as belief,
emotion, and desire, are inherently vague, and fit into no
natural kinds. Moreover, in opposition to the rather bland
functionalism considered briefly at the end of the last
chapter, it might well be argued that our ordinary psychological talk is not just descriptive, but attempts explanations

of our behaviour in terms of reasons and mental states, and as such offers us explanations which are less good than those offered by physics, as they are not systematic, precise, or quantifiable. In the end persons are material beings and their behaviour should be explicable in satisfactory scientific terms, terms which refer to properly clear natural kinds. The mental states of people, on this view, are at most epiphenomena of bodily (material) states on which, to use the jargon, they supervene. In other words, there is no change in a mental state without a change in a physical state, and the assumption is that it is the physical states which are actually causally effective. On this view, our talk of reasons and mental states as the source of our intentional activity will be doubly misleading, because of the implications of such talk regarding freedom, and because of the lack of any precise translations between the terms of neurophysiology and those of folk psychology, which, as already observed, are vague and form no natural kinds. The actual causes of our actions are seen as the law-governed neurophysiological states on which our mental states merely supervene, and which in a highly oblique sense we refer to when we speak of the causes of our actions. So, although we will no doubt continue to use mentalistic folk psychology in our everyday life, such talk could be eliminated for explanatory purposes at least, and such an elimination would result in descriptions of human behaviour that were truer and more precise than the misleading and unsystematic action- and intention-filled descriptions we use in everyday life.

Now, the point about eliminative materialism is that there is barely a shred of hard empirical evidence that it is true and some doubts about its very cogency when we come to consciousness. In other words, the neuroscience predicting and explaining all our bodily behaviour does not exist as yet, nor indeed is it clear what an explanation of consciousness in terms of physical states would look like, how the gap between matter and experience is to be bridged. Such an explanation is clearly needed if the programme is to go through, for the supervenience of conscious states on the physical is part of the account, and must be included if the programme is to be a complete account of human activity. This is not to say that bodily behaviour might not be predicted by some future neuroscience or that people are wrong to investigate the phenomenon of consciousness, even if one may be forgiven some scepticism regarding the very intelligibility of 'explaining' a felt experience, such as pain, in terms of a physical state. But, given that a science predicting bodily behaviour and physicalistically explaining the phenomena of consciousness does not yet exist and that its very form is in doubt, eliminative materialism is hardly a scientific thesis or one demanded by scientific realism. It is a piece of philosophical metaphysics, though perhaps none the worse for that. What is regrettable is the tendency to regard this metaphysical speculation as *the* truly scientific approach to the human person, and to regard those who are bold enough to express reservations about it as reactionary backwoodsmen intent on clinging to scientifically outmoded forms of obscurantism. In fact, the tendency to disparage 'folk' psychology in the name of scientific realism is a good example of the mythology of science in action.

The mythology of science may be regarded as the belief that science and scientific method (and they alone) can provide us with complete and satisfactory explanation of all phenomena. Apart from the limits to the extent of scientific theory already considered, as things stand there is actually in this claim a curious degree of unreflectiveness, amounting almost to a paradox. The point I am making here is that science itself is a human activity and its theories human constructions. As such, they depend for their existence on the existence of human agents and their intentions to develop theories, explore the logic and rationality of those theories, discover the truth about the natural world, manipulate nature, and so on. Scientific theories are not just marks on paper. Unless and until understood and applied by human minds, the inscriptions in which they are embodied are indeed just that; but it is only in being understood and applied that the lifeless inscriptions achieve their full reality as representations, as intelligible systems designed and intended to be understood as representing the natural world. And for such systems to be possible at all there must be a pre-existing human community and language, against the background of which individual human beings can understand particular sounds and inscriptions as scientific theories, and against which new theories they construct can be expressed by them and understood by others. All this is quite apart from the way, stressed by Michael Polanyi and Thomas Kuhn, in which science as an activity depends on the existence in the scientific community of bodies of tacit knowledge; shared, but unformulable assumptions regarding relevance, likelihood, and interest, which are crucial in any piece of scientific problem-solving.

Scientific theories, then, are themselves products of human agency and intentions, and intelligible as scientific theories only if regarded as such. Let us suppose, by way of contrast, that the wind and the rain had carved out something like the letters 'F = ma' on a rock. Would we then want to say that the wind and the rain had inscribed Newton's second law on the rock? It is clear to me that except in a highly derivative sense we would not. In the first place, for the letters 'F = ma' to have the sense they do, a great deal of background interpretation is required, whole systems in which the notions of force, mass, and acceleration have the sense they have. This background is entirely absent in the example. Then, secondly, 'F = ma' only has the force of Newton's law if we could envisage it being intended by its utterer or writer as a description of some aspect of reality, and standing in logical and epistemic relationships to other statements believed by the utterer to be true. Clearly, whatever we might think of the operation of the wind and the rain on rock, the 'formula' cannot be seen in that light. It cannot be seen as obeying the dictates of reason or logical thought in what it does or as intending to do so, or to say what is true. Its activity is entirely a matter of non-rational causal processes, and a cause does not stand in a logical relationship to its effect, nor can such processes be seen as operating under the constraints of rational discourse. They just are, and are not aimed at subserving the requirements of reasoning or rationality, in which, for example, statements are accepted or rejected because they logically follow from or are contradicted by other statements we hold. We would see the marks on the rock as Newton's Second Law only in virtue of the chance coincidence between their form and an inscription of Newton's Second Law, which was produced with the appropriate intentions within a whole system of human purposes and communication. And the same goes for computers endlessly printing out scientific formulae and results. These print-outs exist as *intelligibilia* only by being seen within the systems of human knowledge, rationality, and activity from which they stem.

If scientific theories, then, derive their meaning and force from systems of human agency and intention, it would indeed be paradoxical if one of their achievements was to be regarded as the denial of the ultimate reality or effectiveness of human agency and intentionality. The connection between this point and the completeness of scientific explanations is as follows. The materialist might admit the points just made about human agency and intentionality and the need for the background of a community of such agents for any rational science, but he is then faced with the problem that science has so far been unable to give an explanation of these and other phenomena of consciousness in terms of neurophysics, let alone deliver a full account of bodily activity. Science, then, is not a complete account of the natural world, given that these phenomena exist as part of the natural world, and it is surely premature, to say the least, to accept a metaphysical view whose cogency depends on the explanation of human mental and physical activity in neuro-physiological terms.

Let us suppose that we did succeed in showing that consciousness, agency, and intentionality are no more than epiphenomena supervening on a basically unconscious and non-rational world of matter. In doing that, would we not be casting some doubt on the reality or significance or rationality of science itself, as a product of the human mind? We would be saying that at bottom its inscriptions were no different from any other actions of atoms upon atoms, just like the effects of the wind and the rain on the rocks. Would we not thus be depriving science of its significance as a realm of intelligible structures and as an arena for and product of the exercise of human reason and agency? There is more than a suspicion that the effect of a science which denies the ultimate significance of human agency, except as an epiphen-omenon of basically unconscious, non-rational, and unin-tending particles, would be to deprive science itself and its theories of their meaning and import, as intended and intelligible descriptions of the natural world, and as a coher-ent system of logical, rational discourse. This is at least in part because the reductionist neuroscience would be saying that the actual reason I uttered the sounds 'F = ma' on a given occasion was because of the causal effects of certain impulses in my brain, and not because I was *rationally* convinced of the cogency of Newton's second law or found it *appropriate* to utter it on a given occasion. It is, in other words, far from clear that any purely causal account of human activity could satisfactorily explain the normative aspects of rational discourse without undermining their normativity. A science which asserts that it is rational to believe that all human activity is ultimately due to the operations of non-rational laws and processes is in imminent danger of undermining its own pretensions to rational ac-ceptability.

If one of the functions of a mythology is to provide a group or a culture with an all-encompassing view of and attitude to the world, there is a strong case for seeing science as the dominant mythology of our time, sustaining metaphysical attitudes such as reductionism and materialism, particularly about the mental and the human, and rationalistic approaches to problems of human living. But science, properly understood, does not justify such attitudes or approaches. Nor does its own practice depend on an espousal of them. In this respect, science differs from other mythologies. The practice of the Christian religion, say, can hardly survive loss of belief in the Christian deity; the practice of Marxism looks increasingly bankrupt, given the evident falsity of Marxist dogma. But no analogous bad faith need be involved in the practice of science without a commitment to reductionism or materialism. Indeed, there is a strong case for thinking that the health of our culture increasingly depends on drawing a clear distinction between science, as it should be understood and practised, and Comtean progressivism, with its tendency to downgrade the significance of lived human experience and to hack away at the delicately balanced systems of value and meaning we have evolved over time for living our lives.

Myths and Science

If science itself can take on some of the characteristics of a mythology, it is also true that science, being part of the culture produced by human beings, cannot remain immune from other cultural and ideological influences. Science may be influenced by myths prevalent at a given time and place, and its practitioners respond in their research and theories to the interests and desires of the ruling groups in society. According to proponents of what has come to be known as the strong programme in the sociology of science, scientific knowledge itself (or, more properly, what is called scientific knowledge at any given time) can and should be given a naturalistic explanation in terms of events and pressures external to science itself.

There can be no doubt that scientists, individually and collectively, can be influenced by extra-scientific forces, including non-scientific myths. We have already drawn attention to Kepler's Pythagorean background and hopes. Newton was strongly influenced in his thought not only by his voluminous researches into biblical apocrypha, but also by the hope that his natural philosophy (or physics) would impress on people the necessity for a divine intelligence to impart to material particles their initial order and motion. He also saw the existence of God as necessary to sustain the notions of absolute space and time. (Here defenders of sociological analyses of scientific theory might not unfairly comment that Newton's physical researches did not provide him with any unequivocal need to postulate absolute space and time; was Newton perhaps predisposed to this idea by his pre-existing concept of God's existence and omnipresence?) It is also commonly stated that both the theory of evolution and quantum physics are to be seen as products of the societies in which they developed, devil-take-the-hindmost *laissez-faire* capitalism and the uncertain, crisis-ridden Weimar Republic, respectively. And no doubt some psychohistorians will say that Einstein's status as a displaced person might have made him more open to relativistic speculation than had he been a confident citizen in a settled milieu, though I suppose others might give precisely the same fact and his subsequent problems in Germany as an explanation of his hostility to indeterminism and the principle of uncertainty.

There can be no doubt that scientific ideas come from many sources, scientific and extra-scientific. In view of this, it would be surprising if one could not sometimes discern *zeitgeistliche* aspects to scientific theorizing. Only a naïve

Baconian inductivist, intent on cleansing science of all elements of presupposition, would be inclined to deny the impact of non-scientific influences on scientific thought. Clearly some of these influences will come from currents of thought prevalent at a given time and place (although influential scientists might be seen as reacting to their time as much as conforming to it). None of this should dispose us to rush into a denial of the distinction between the context of discovery and the context of justification.

In the first place, even in the process of formulating his theory, a scientist will be tempering his non-scientific inspiration. The theory he proposes will qualify as properly scientific only because it is an answer to an existing scientific problem and is cast in a recognizably scientific form. Kepler might have believed in number as the essence of the universe but, unlike most numerological speculation, his theory was scientific and assessed as such because it addressed the already existing scientific problem of the motions of the planets and did so in a way which took into account and appeared to explain the relevant observational data. Equally, to say that some form of belief in the survival of the fittest was dear to the heart of every Victorian capitalist does nothing to explain the nature of Darwin's scientific achievement in writing *The Origin of Species* (which did not, incidentally, provide any support whatever for the equally Victorian belief in progress, even in the biological realm, a suggestion which Darwin consistently and properly resisted). Darwin's work was taken seriously as biology, and continues to be taken seriously in a period with a quite different ethos, because of its fruitfulness and success in dealing with masses of biological data at various levels, and it clearly would not have had the biological impact it did had Darwin himself not addressed himself in his work to the specifically biological issue of speciation and amassed a wealth of relevant data.

Then, secondly, even where a theory has been cast in a

scientifically acceptable form, it success or failure scientifically is not directly related to its conformity or otherwise to the non-scientific spirit of the age (assuming for the moment that one is entitled to speak of such a thing). Not only do we have the example of a theory like Darwin's whose success and scientific viability persists through changes of age, but there are notable examples of theories being rejected despite their conformity to the spirit of the age. The rise of capitalism may well have been an environment friendly to the atomistic and individualistic world-view of Cartesian mechanics, but even while the new economic system was conquering new fields, scientists were compelled to temper the extreme individualism of Cartesianism with the more holistic and communitarian concepts of action at a distance, and even of the inter-relatedness and inter-dependence of the whole universe.

What, then, are we to make of the claim that theories are accepted (rather than acceptable) because they fit in with the spirit of the age; that, for example, Newtonian physics, by virtue of its divinely based natural order, was attractive to those who would think of civil society and its orders as divinely established; that quantum theory seemed more acceptable than it otherwise would have done because of the Spenglarian ethos of doom and mystery reigning in post-1918 Germany? We should note first that, as with Kuhn, those who propose such naturalistic and denigratory analyses of scientific rationality rarely turn their analytical weapons on themselves; their claims, apparently, escape reduction into the relativities of the sociological and the historical. Quite apart from this, we must continue to insist that science itself has as one of its internal constraints the requirement that its theories fit with nature. Flexible and somewhat uncertain as this requirement may at times be in practice, it nevertheless does represent something which cannot be avoided by appeals to the spirit of the time. Given, too, the essential openness and competitiveness of the scientific community and the requirement of repeatability of observations and results, it is going to be very difficult for proponents of a theory to hide for long the existence of counter-evidence or lack of empirical fit. If a ruling theory does not fit the facts, someone in the end is going to blow the whistle, if only because he will make his reputation by doing so successfully. As already argued earlier, empirical fit of nature and theory is not enough to determine theory-choice uniquely. But striking lack of fit will rule out some theories altogether, and this fact allows us to insist on a degree of scientific rationality, or context of justification, which cannot be collapsed into history or sociology. Despite all his prestige and the full backing of an oppressive and dictatorial ideological state apparatus, Lysenko was not able to make his wheat grow. Even in that darkest of environments, geneticists recognized this fact, and some indeed were prepared to suffer for this recognition.

Some will say that in this section I have conceded too much to the strong programme of the sociology of science, and that the very stability of science throughout the world and across ideological boundaries, referred to in the introductory chapter, shows the autonomy of science from other social and ideological factors. In reply to this objection, it might be questioned whether there might not be ideological factors which themselves cross certain ideological boundaries, as between Marxism and Liberalism, for example. Perhaps the type of reductive materialism considered in the last section might be an example of such a thing. Whether this is so or not, however, what I am suggesting here is that even on the assumption that there are extra-scientific factors influencing scientific thought and practice, we can and must preserve some sense that science, by the practical and theoretical obligation which its theories have to fit nature, is not completely circumscribed by these factors. To the contrary, it surely is the case that the demand for a fit with nature gives to scientific objectivity a hard edge which is lacking in other areas of human activity, such as the arts and literature, where acceptability is defined entirely in terms of human response. Indeed, the key point to grasp here is that scientific rationality is constituted not by the provenance of ideas, but by their criticism and testing. In science, it does not matter where ideas come from, so long as they are then tested in an open spirit.

It would surely be wrong to deny that there are any extraneous, non-scientific influences on the history and development of science. Even if, as the last paragraph suggests, it is less easy to see narrowly ideological influence as important in science now than it may have been in earlier times, the scientific community itself is not always as open and responsive to critical or dissenting voices as it should be. There is also the influence of technology and its imperatives on science, which we will consider in the next section. The short answer to the point about technology is that its imperatives cannot by themselves distort scientific truth. To the contrary, discovering the success or failure of a piece of technology is a prime example of the discipline of nature to which scientists have to submit. On the lack of openness in the scientific community, one can certainly notice manifesta- tions of closed-mindedness, such as the length of time it took for Fourier or Faraday to gain acceptance for their unfash- ionable ideas. That they did eventually gain acceptance for them is not in itself a complete vindication of the rationality and openness of the scientific community. There might, after all, have been other potential Fouriers or Faradays who never broke through the barriers put up by the scientific establishment of the time. On the other hand, none of this detracts from the significance of the fact that in science the ultimate dissenting voice is nature itself, and that is a voice which even an entrenched scientific establishment cannot silence for ever. In fact, despite failures of openness and authoritarian tendencies which undoubtedly do exist within the scientific community—and which have been documented by its critics almost *ad nauseam*—we must continue to insist that honesty in experimentation, openness to rival views, and criticism of established views are values honoured in science, and to an extent enshrined in its institutions. As already mentioned, experiments and observations must be repeatable, and reputations can be made by well-grounded criticisms of established theories.[1] Moreover, in most fields of science, currently at least, there is no overall agreement on a single correct theory, as we see with, for example, conflicting interpretations of quantum theory or of biological evolution. All this, of course, contributes to an atmosphere of free discussion in science and militates against the domination of scientific thought-processes by ideological or self-interested forces, whether they are within or without science itself.

Science and Technology

Why did the ancient Greeks not achieve more in the field of technology? This most ingenious and rational of people tended actually to despise the merely mechanical or banausic, despite their interest in scientific and cosmological speculation. Maybe, as Jasper Griffin has suggested, they esteemed aesthetic values higher than the technical.[2] If this was so, though, it by no means implied that they did not prize scientific knowledge, knowledge of causes, knowledge of nature. But these things were valued for their own sake and not for the manipulative advantages they might bring with them.

The modern world, by contrast is largely governed by the

[1] It is important to insist that criticisms and alternative points of view have to be empirically well-grounded before there is any duty on the part of the scientific community to attend seriously to them. Scientific rationality is not impugned by the failure of scientists to take flat-earthers, creationists, or believers in mystic lines of force seriously.

[2] In J. Boardman, J. Griffin, and O. Murray (eds.), *The Oxford History of the Classical World* (Oxford University Press, 1986), p. 7.

concept of technical progress, which sweeps all other considerations before it. Griffin points out that the beauty of the riders on the Parthenon frieze depends on the absence of stirrups and hints that realization of this beauty might have impeded experimentation with aids to more efficient methods of riding on the part of the Greeks. One cannot imagine the modern world holding up technical progress on aesthetic grounds. Indeed, if the so-called entertainment industry is a guide, technology has come to dominate aesthetics even in the realm of art, to say nothing of its effect on architecture and design generally, thus fulfilling a thousand times a day Goethe's sombre warning that technology in alliance with bad taste is the enemy of art most to be feared.

We noted earlier how Bacon saw science as a significant way of contributing to the relief of man's estate by giving power over nature. Undoubtedly the rise of modern science was accompanied by a growth in science-based technology, and the prestige of modern science derives in large part from the success of technological innovation. Undoubtedly, too, scientific research is often skewed towards technological hopes, something which becomes more and more significant with the increasing cost of scientific research.

In the light of the contemporary intertwining of science and technology, Karl Popper, writing in the mid-1950s, comes across as firmly Hellenic and un-Baconian in his approach:

The nuclear bomb (and possibly also the so-called 'peaceful use of atomic energy' whose consequences may be even worse in the long run) have, I think, shown us the shallowness of the worship of science as an 'instrument' of our 'command over nature' or the 'control of our physical environment': it has shown us that this command, this control, is apt to be self-defeating, and apt to enslave us rather than to make us free—if it does not do away with us altogether. And while knowledge is worth dying for, power is not.[3]

---

[3] *Realism and the Aim of Science* (Hutchinson, London, 1983), p. 260.

While some sections of the ecological and environmental movement might be surprised to receive apparent support from such a source, these views are consistent with Popper's philosophy as a whole, in so far as they derive from a strong sense of our ignorance about nature and about the effects our interventions might have on nature. As we shall see shortly, realization of the extent of our ignorance about the natural world poses critical problems for our thinking about technology, problems which are often not fully appreciated by either its defenders or its critics.

Before moving on to this, however, it is worth pointing out once more that whatever may be the moral and political faults of engineers and technologists, and of the institutions, politicians, and communities who rely on them, technology can have no tendency to undermine the cognitive claims of science. Indeed, somewhat against Popper, the success or failure of the technological applications of theories are highly relevant to assessing their truth or falsity. It is also worth emphasizing the obvious, but often overlooked point that it is not scientists *qua* scientists who decide that a society should order or encourage such things as atomic bombs or nuclear power or genetic engineering, however much individual scientists may press political authorities or commercial interests to sponsor their work. But this separation of the scientific from its social applications raises the question of the assessment and evaluation of a particular piece of technology.

The introduction of a piece of new technology is an experimental matter. The entrepreneurial engineer moving into the market does so usually in order to solve some perceived problem, such as satisfying a need or creating a desire in people, or improving or correcting the unwanted effects of some existing device. He may, of course, be wrong. There may be no gap his device fills, or he may fail to pick out a desire anyone has even potentially, or his device may create more problems than it solves. What is certain is that, like all human actions, a new answer to an old problem will bring with it unforeseen and unforeseeable consequences, and some of these will be seen as new problems, themselves requiring further new solutions, which will bring yet more problems in their train, and so on and so on.

Part of the reason for the unforeseeability of the consequences of technological innovations is something inherent in any new piece of scientific theorizing or experimentation. That is, our ignorance of nature is such that until we test our theory or carry out our experiment, we just do not know what the result will be. And even a laboratory simulation of real-life conditions has an element of risk or uncertainty about it. The laboratory simulation is one which fixes on certain factors as being relevant in the real-life situation, and it may just be that some factor not chosen for representation in the laboratory simulation turns out to be relevant, often to the detriment of the real-life application of the technology. This is not, of course, an argument against laboratory testing of new technologies or theories, but a caveat about the inherent limitations of such testing. Many disasters are prevented by such things, but not all. So our ignorance of nature and its effects means that, however careful we are in devising a theory or piece of technology, we must always accept that there may be unforeseeable outcomes in testing the theory or applying the technology.

But a piece of technology is not just an intervention in nature. It is also an intervention in the human, social world. In this sense, there will always be an entrepreneurial dimension to engineering. Will our new technology catch on with the public? And, if it does, what will be its effects more generally? Here we reach another dimension of uncertainty, and ignorance, at least as critical as our ignorance of the natural world. For the effects of social actions are subject to multiple uncertainties, due both to the extent of their effects and the way those effects depend in part on the unpredictable responses of individuals. The unpredictability of response is doubtless magnified when we are dealing with a piece of new technology. To take a simple example, not only did no one predict the way computers would catch on with the general public in Britain, but it is extremely doubtful that anyone in advance of the event would even have taken seriously the thought that video games would have been so significant an aspect of the home computer market.

Uncertainty about technical effects and developments of new technology, combined with uncertainty about the social effects of technical innovations, leads us naturally to the conclusion that, in the words of Milton Mueller,

the form taken by a technological system is not a *design* but an *evolutionary trajectory*. The trajectory is defined as its operations adjust to the specific social, geographic, economic and political characteristics of its environment, and overcome the technical problems posed by its growth and its competition with other technologies. Anyone familiar with the history of a large-scale technological system knows that an attempt to implement the simplest idea can create vast numbers of unforeseen problems. It is the process of solving these problems—not a preconceived design—that gives the system its shape.[4]

An evolutionary system is one which develops without a guiding reason behind it, by means of leaps in the dark moderated by environmental pressures. Due to our ignorance of the natural world, and of the outcome of our predictions, science itself is at least partially an evolutionary system, in this sense, and so is technology.

From the evolutionary and unpredictable nature of technology, two important conclusions follow. First, the attempt to predict the direction or outcome of a particular technological innovation in advance is bound to be uncertain. This cuts against both sides in many debates about technology: against those who would advocate the development of a particular technology and against those who would seek to

[4] M. Mueller, 'Technology out of Control', *Critical Review*, 1. 4, 1987, pp. 24–40, p. 32.

control it, in so far as their positions are based on the certainty of their predictions about the effects and future development of the technology. In so far as such predictions cannot be made with any certainty, technological forecasting is a largely bogus and fraudulent enterprise. The bogusness increases when we take into account not only the future impact of existing technologies, but also the impact of as yet uninvented technologies, many of which will naturally interfere in far-reaching and unsuspected ways with the evolution of existing technologies. We cannot know what will be invented, even in the very near future, for if we could, we cound invent it now. Nor can we have in advance any idea of the impact of a new invention, as the impact of the electronic computer shows. Even now, with several years experience of the thing, we would be hard put to assess its impact on our present lives, because the relevant information is distributed through so many parts of our lives, and dispersed through so many different sectors of our society, administration, technology, medicine, finance, the leisure industries, the mass media, and so on. And if we cannot yet retrospectively assess its impact up to the present, what chance is there of grounding forecasts of its future impact in any rational way? This type of futurology contrasts with scientific prediction; with technological forecasting we are always in a state of multiple ignorance regarding both initial conditions and relevant laws.

If technological forecasting is thus inherently bogus, many of the characteristic stances of advocates and opponents of specific types of technology are undermined, for these stances are often based on future projections about how a technology will develop and its social effects. Seen in this light, attempts by governments to control and direct technology begin to look bogus too, and likely to have the opposite effects to those intended. Quite apart from the potential for corruption opened by such controls, they are likely to deprive citizens of the freedom to improve their standards of living by choosing the products of better technologies, and to strangle native industries by making them stick with outdated and eventually unprofitable goods, leading in the end to the eventual decline of those industries in the world market, and at home as well once foreign competition gets a foothold.

On the other hand, admitting the spuriousness of technological forecasting does not mean that we should not monitor the effects of technology closely. Indeed, the very reason for the failure of technological forecasting—our profound ignorance of the future evolution and effects of technology—makes the monitoring of its effects that much more pressing. In this context, it is surely significant that it is precisely those societies which go in most for the attempt centrally to control technological development which suffer the worst pollution of their environment by technology. If we want the evolution of technology to proceed with the least cost humanly and environmentally, we should concentrate our efforts on a sensitive monitoring of its effects, in order to gain a clear understanding of its unforeseen consequences and of the new problems it sets us to solve. And if the scientific community is, ideally, a microcosm of an open society, in that criticisms and ideas are permissible from any quarter, it may well follow that a politically open society is the best arena for an acceptable evolution of technology, for in such a society, the citizens will be free to choose among competing technologies, and to draw public attention to and, if necessary, legislate against the unacceptable effects of existing technologies.

In a free society, citizens individually or collectively will be able to resist and question the so-called technological imperative, by refusing technologies they find morally or in some other way objectionable, just as Griffin's Greeks might have refused to entertain the stirrup. Absence of probably counter-productive central planning of technological development in no way implies lack of controls, social or legal, against undesirable technological results. The moral for healthy technological development seems to be not to interfere with its spontaneous evolution by attempting to control or plan its development by centrally directing and curbing technological entrepreneurship on the basis of technological forecasting, but to be severely critical of any adverse effects. Such a policy would, in effect, be simply a version of Popperian falsificationism applied to technology, rather than to theoretical science.

What, though, of the Aristotelian–Popperian hostility to the instrumental uses of science? Like much else in which there is a grain of truth, such hostility is something which can surely be overdone. It is true that there is a sense in which the desire to know the truth about nature is worthwhile in itself, independent of any instrumental side-effects. Indeed, part of understanding who and what we are will be dependent on our knowledge of the natural world. On the other hand, the 'relief of man's estate' is a noble and worthwhile goal too, and one which should not be disparaged as merely instrumental.

Science and Value

There are those who question whether science has, in fact, contributed to the relief of man's estate, and those, too, who would argue that its net effect has actually been the opposite, blaming on 'Western' science many of the ills that afflict the world today. Perhaps, as for Popper, this resistance to science begins with a distaste for the very idea of controlling nature, combined with a vivid sense of the dangers inherent in the attempt. Against this, though, it must be said that there is nothing in science itself which implies that one should intervene recklessly in nature and disturb the natural order, without carefully monitoring the effects of what one does. Nor is it science or scientists who decree that so much effort and money should be devoted to military research, and there is something disingenuous in citizens of a democracy blaming scientists for working on military projects, if that is where their governments have decreed money is to be spent. But a more radical form of criticism of the technological applications of science, and one we have not so far touched on comes from the claim that there is something inherent in the concept of science itself, as a value-free investigation into aspects of the natural world, which makes it of necessity a destructive and malign influence on society. The claim is that there is something inherent in the methods of science which make its practitioners peculiarly prone to go in for, say, weapons research, as opposed to solving the world food problem and devoting themselves to working out ways in which we can all collectively decide on a better, wiser way of life for everyone on the planet. This, at any rate, is the claim of Nicholas Maxwell, in his influential book *From Knowledge to Wisdom*.[5]

As this sort of criticism of science is common, and not confined to Maxwell (who, indeed, argues his case with more understanding of science and philosophy than many who write on these topics), perhaps one might begin by suggesting that knee-jerk reactions against the 'arms race' and in favour of 'solving' the world food problem are themselves open to question. If, as is certainly arguable, it is the case that a war in Europe and the spread of Soviet Communism have been prevented by weapons research and the 'military-industrial complex', then so much the better for weapons research and the 'military-industrial complex'. (It is surely better to have scientists working away in laboratories and the 'military-industrial complex' producing weapons which are not used, than another European war and gulags in Paris and

---

[5] Basil Blackwell, Oxford, 1984. See especially p. 1 for Maxwell's version of the modern world and the implication that science and the academy are in 'a major way' to blame for our failure to respond 'sanely and rationally' to the global problems facing us.

London and Munich and Rome.) On the world food problem, it is perfectly clear that *science* has (at some cost to
landscape) provided us with the means to feed the total
population of the world several times over.[6] So it is a bit
unfair to blame science for our failure to eradicate starvation;
and we should certainly acknowledge the remarkable strides
science has made in the eradication of infant mortality,
cholera, smallpox, and many other earlier scourges of humanity.

Many will find the tone of the last paragraph unacceptably
robust, but it should at least suggest that certain assumptions
often taken for granted in attacks on science are not themselves indubitable. However, Maxwell's criticisms of science, though undoubtedly coloured by his political perceptions, do not depend on them. Rather they derive from the
very nature of science as an enquiry which prescinds from
questions of value. Its very neutrality in enquiry counts
against it morally because it means that its practitioners are
able to engage in research and technological applications of
their research without having to ask themselves moral questions about the value or likely social effects of their work.
Against the claim of Sir Andrew Huxley, that, 'in the long
run the value of science depends entirely on its conclusions
being independent of wishes and fears about their practical
application',[7] Maxwell urges that knowledge is valuable only
to the extent that it is knowledge of valuable truth, that is
truth which helps us collectively to 'live life lovingly', as he
puts it. Against Dr Bronowski's position, expounded in his
book *Science and Human Values*, that we cannot blame
science for what happened at Hiroshima and Nagasaki,
Maxwell says that 'our task, in engaging in rational enquiry
is to see, participate in, and help to grow what is significant

---

[6] Maxwell recognizes that we have the capacity to provide food enough for
everyone to get enough to eat (see his p. 1). But where does he suppose this
capacity comes from?

[7] Quoted by Maxwell, in *From Knowledge to Wisdom*, p. 133.

and of value in existence in the cosmos'.[8] Hiroshima and Nagasaki represent failures in that task, failures actually made possible in large measure because of the dissociation scientists characteristically make and are trained to make between scientific knowledge (value-free) and its applications (value-laden). Maxwell, by contrast, would make the assessment of the aims of scientific research internal to the scientific activity itself, and not something external to and separable from what is regarded as the purely scientific stratum of a scientist's work. He, like many other critics of science, insists that the implicit separation of the internal knowledge-producing aspects of science from enquiries into the use and value of the knowledge thus acquired provide all too easy an alibi for scientists to engage on harmful and destructive projects, and may indeed actually encourage the growth of such projects, in so far as those who work on them are, in the spirit of Huxley's pronouncement, trained to look at them in isolation from their wider social and moral effects.

Maxwell's criticisms of scientific practice clearly have connections with contemporary feminist and other radical criticisms of science, which see modern science as more concerned with the qualitative analysis, control, and profit-oriented exploitation of nature than with understanding nature in holistic terms. Like Maxwell, a feminist writer such as Sandra Harding[9] will maintain the inseparability of questions of values from scientific activity, claiming even that emancipatory values are actually conducive to scientific objectivity. Masculine-based 'oppressive' mechanist metaphors, according to Harding, may have been beneficial to science in the past, but nowadays it is 'only coercive values—racism, classism, sexism—that deteriorate objectivity'.[10] Like Maxwell, Harding believes that participatory values will actually produce better science than the standard

[8] Maxwell, in *From Knowledge to Wisdom*, p. 8.
[9] *The Science Question in Feminism* (Cornell University Press, Ithaca, 1986).
[10] Ibid., p. 249.

empiricist methodologies, such as the one outlined and defended in the first six chapters of this book.

Maxwell argues in his book that empiricist methodologies make it impossible to understand the progress of science and, by implication, that such methodologies will hold up scientific progress. This is because empiricism refuses to recognize that ultimately the universe is comprehensible in ways which will be revealed as we go along, and as we recognize ourselves as part of the evolution of nature towards states of greater value and greater love.

According to Maxwell—and here, while drawing the opposite conclusions, we would agree—empiricism is unable to explain how science can reveal the deep structure of the world, and has to interpret science largely in terms of its experimental and predictive results. Empiricism thus deprives science of its intellectual richness as a loving attempt to grasp not just empirical results but the architectural grandeur of the universe. Maxwell's vision, and one on which any 'participatory' theory of science must ultimately rest, is of the human mind (at least when not perverted by false values) as attuned to nature in its essence, and expressing its love for nature in doing science.

I will not repeat here the earlier discussions of empiricism and realism. What I will suggest, though, is that Maxwell and Harding are mistaken in their shared view that only an 'emancipatory' science will be truly objective, and that Maxwell's attempt to conflate the assessment of aims in science with the pursuit of knowledge is likely to be damaging to both activities. One can make these points without saying either that we should not assess the direction of scientific research on moral or political grounds, or that scientists themselves should not consider such questions. What we need to assess, though, is the correctness of Huxley's claim that the validity of the results of science depends in the end on the independence of its conclusions from wishes and fears about their practical application.

In the first place, it must be perfectly obvious, despite Harding and Maxwell, that the backing of a given set of political or metaphysical values is neither necessary nor sufficient for scientific truth. The anti-Darwinists in the Soviet Union in the 1940s did not achieve scientific truth, despite being motivated in their theories by correct 'progressive' values. Nor is it clear that the realist Einstein, lauded as such by Maxwell, was correct in his opposition to the largely instrumentalist and non-realist proponents of the Copenhagen interpretation of quantum mechanics. Indeed, one can say quite generally that such values as political emancipation (however conceived) and participatory realism are not achieved in advance of knowledge of the relevant aspects of nature. We cannot, as Harding and Maxwell appear to want us to do, assume in our scientific work one version of a specific value and then expect that nature is obligingly going to fit it. In so far as nature impinges on the realization of our values, we may have to modify either the values themselves or our ideas about how they are to be applied. Huxley's point here is correct: people respect the findings of science precisely because they can be separated from sets of values, political or metaphysical, and provide a realistic framework for the realization of our ethical and political projects.

The reason for this is easy to see. Science aims at the discovery of causes and regularities in the physical world, in a world, that is, with an existence apart from us and our concerns, and not necessarily attuned to them. Science thus necessarily prescinds from the way things appear to us, matter to us, and have significance for us. It may indeed show us that the way things seem to us and matter to us are not revelatory of the way nature actually works, as colour, for example, is downgraded in scientific terms to being a merely secondary quality and not a causally fundamental property of matter. But this knowledge can do nothing to diminish the importance of colour in our everyday existence or the emotional significance colour has for us. It is precisely because science abstracts from the appearances and the lived significance of things that, as we noted at the outset of the book, the findings of science will be acceptable to scientists from very different political and religious backgrounds.

The great strength of science intellectually and practically is its ability to deliver accounts of the world which abstract from human life and our concerns, concerns which we express and develop in and through our values and ideologies. It is its strength intellectually because it is to the credit of science and scientists and to the societies which support them that human beings are capable of forming or attempting to form conceptions of the world as it is, rather than as it matters to them. This is a testament both to our intellectual and to our moral nature: to be able to find out something of how the real world is and to submit to the disciplines necessary to make this possible. And it is science's strength, practically speaking, because, in Bacon's terminology, those who best obey nature will in the end best command her.

Science, then, in order to serve its aim of describing and explaining the world as it is apart from us, stands at one remove from human response. As one physicist remarked, physics is about how atoms appear to atoms. For this reason one has to be suspicious of Maxwell's aim of conflating the descriptive and explanatory role of science with consideration of questions of value, including those relating to the value of particular types of scientific enquiry. In fact, the tendency to generalization and reduction, which is appropriate to the aims of science, is ill suited to the making and assessing of judgements of value.

One is reminded here of the debate in the early 1960s between C. P. Snow and Dr Leavis on the existence or non-existence of the 'two' cultures. What had particularly enraged Leavis about Snow was Snow's bluff and bland optimism about the power of a rationalistically and scientifically directed world to solve problems of strife and starvation. Leavis argued in response that the scientific approach to social problems eliminated human life: 'the complexities it reduces them to are mechanical, or treatable mechanically—it hates the organic, and its simplifications kill.'[11]

Without even pausing to consider the fruits of a rationalist-cum-scientific approach to social problems (such as modern architecture and town planning), it is clear that Leavis touches on something important here, something which actually differentiates the methods of science from those of enquiries more directly concerned with questions of human life and value, such as reflection on morality and on works of art and literature. For in the latter type of enquiry, what is crucial is the observation of the nuance and detail of how things appear to us and affect us and matter to us, how they mesh or fail to mesh with the fabric of human value, and what is required on the part of the commentator is a mature and lived sensitivity to such things. As Leavis puts it (while inveighing against the idea of a separate literary culture alongside a scientific one): 'the judgements the literary critic is concerned with are judgements about *life*. What the critical discipline is concerned with is relevance and precision in making them.'[12] The point is that science, being about the nature of things, rather than about their aspects for us, about how they affect and might affect us, is an ill preparation for the required sort of relevance and precision.

It is perhaps no coincidence that Snow and Maxwell, for all their differences, both believe that problems of value and ideology would be solved if only a proper sort of scientific rationality were in place. In this common assumption, they fail to see that scientific rationality, however widely conceived, is always going to be tangential to how people feel

[11] F. R. Leavis, *Nor Shall My Sword* (Chatto & Windus, London, 1972), p. 207.
[12] *Nor Shall My Sword*, p. 97.

and react and to the sources of their values. Scientific rationality is a tool for discovering general and increasingly abstract truths about nature, rather than about how things appear to us and affect us,[13] and it is therefore unsuited to examining the lived texture of human experience, an area in which there may be no general theoretical understanding to be had, and where feeling and reaction are of the essence.

What has just been said is not, nor is it intended to be, any denigration of science or the scientific attitude. It is in science and by means of its methods that we find out essential truths about the nature of the world. Science also provides us with the best model we have for honesty and integrity in enquiry. We turn our back on science and its virtues at our peril. As Goethe put it, arguing against those beautiful souls who would retreat entirely into artistic inwardness, man knows himself only in so far as he knows the world. One of the things we must take into account in considerations of meaning and value is the actual nature of the natural world, our roots and origins in that world, and our precarious relationship to it, and these are all things on which we need science to enlighten us. On the other hand, as Goethe knew as well as anyone, it is not in science that the observer of the world explores the meanings and potential meanings things in the world can have for him or establishes just how he should relate to the world or to his fellows. Science's very strength is the way it prescinds from questions of value in an attempt to adduce observer-independent truths about a world independent of us. But it is just this feature of science which makes it unable to tackle questions of value head on and, as I suggested in the first chapter of this book, these are the most important questions. The advocacy of a science which conflates questions of fact and questions of value will not produce either good science or better reflection on

[13] On the tension between scientific approaches to phenomena and our lived experience of those phenomena, Thomas Nagel's *The View from Nowhere* (Oxford University Press, 1986) is most suggestive.

questions of value. In its failure to recognize the essential differences between science and the humanities and, by implication, the essential nature of science as aiming at objective and value-free knowledge of the natural world, such an approach is likely to subvert both true science and a proper understanding of the source of value in the lived experience of generations of human beings.

Suggestions for Further Reading

A standard introduction to the philosophy of science, now somewhat old, but still very useful, is Carl Hempel's *Philosophy of Natural Science* (Prentice Hall, Englewood Cliffs, 1966). Two other classic works outlining standard views of scientific methodology are Hempel's *Aspects of Scientific Explanation* (Free Press, New York, 1965) and Ernest Nagel's *The Structure of Science* (Harcourt, Brace and World, New York, 1961). A more recent and very stimulating survey of the current situation in many central issues in the philosophy of science is Ian Hacking's *Representing and Intervening* (Cambridge University Press, 1983).

On the inductive method, Bacon himself is well worth reading (*Novum Organum*). J. S. Mill's *A System of Logic* (1843 and subsequent editions) is more systematic and, in a way, more inductive. Popper makes trenchant criticisms of the method in his *Conjectures and Refutations* (Routledge & Kegan Paul, London, 1962), which is also an excellent introduction to his own thought. Popper's *Logic of Scientific Discovery* (Hutchinson, London, 1959) and *Objective Knowledge* (Oxford University Press, 1972) should also be accessible to readers of this book. Wesley Salmon defends a somewhat Bayesian approach to scientific theorizing in his *The Foundations of Scientific Inference* (University of Pittsburgh Press, 1976). John Watkins in *Science and Scepticism* (Hutchinson, London, 1984) deals head on with problems for science raised by Humean scepticism and attempts a sophisticated Popperian response.

On the demarcation of science from non-science and questions relating to the falsifiability of scientific theory *Criticism of the Growth of Knowledge*, edited by Imre Lakatos and Alan Musgrave (Cambridge University Press, 1970), is invaluable. Kuhn's *The Structure of Scientific Revolutions* (University of Chicago Press, 1962) is indispensable and highly readable. Paul Feyerabend's views are set out in his *Against Method* (New Left Books, London, 1975). Peter Munz discusses Kuhn from a historical point of view in *Our Knowledge of the Growth of Knowledge* (Routledge & Kegan Paul, London, 1985). Dudley Shapere is also critical of Kuhn's history in 'The Structure of Scientific Revolutions', *Philosophical Review*, 1964, pp. 383–94, and 'Meaning and Scientific Change', in R. Colodny's *Mind and Cosmos* (University of Pittsburgh Press, 1966). The latter essay is reprinted in a useful collection of articles, *Scientific Revolutions*, edited by Ian Hacking (Oxford University Press, 1982). In this context, mention should be made of Larry Laudan's *Progress and its Problems* (Routledge & Kegan Paul, London, 1977) and W. Newton-Smith, *The Rationality of Science* (Routledge & Kegan Paul, London, 1981).

On questions related to observation and theory, apart from works already mentioned, N. R. Hanson's *Patterns of Discovery* (Cambridge University Press, 1958) is a classic. Also worth reading is Feyerabend's 'Problems of Empiricism' in R. Colodny's *Beyond the Edge of Certainty* (Prentice Hall, Englewood Cliffs, 1965). The notion of the underdetermination of theory by data has been extensively discussed by W. V. Quine, especially in *Word and Object* (MIT Press, Cambridge, Mass., 1960) and *Ontological Relativity* (Columbia University Press, 1969). Quine also discusses the notion of natural kinds in the latter volume. A more essentialist handling of that question is to be found in Hilary Putnam's 'The Meaning of "Meaning"' in his *Mind, Language and Reality* (Cambridge University Press, 1979), in Saul Kripke's *Naming and Necessity* (Basil Blackwell, Oxford,

1980), and in David Wiggins's *Sameness and Substance* (Basil Blackwell, Oxford, 1980).

On scientific realism, the best place to start is Bas van Fraassen's *The Scientific Image* (Clarendon Press, Oxford, 1980). Van Fraassen's work has been discussed in *Images of Science*, edited by Paul Churchland and Clifford Hooker (Chicago University Press, 1985). Apart from Hacking's *Representing and Intervening* and Nancy Cartwright's *How the Laws of Physics Lie* (Clarendon Press, Oxford, 1983), mention should be made of Nelson Goodman's *Ways of Worldmaking* (Harvester, Hassocks, 1978) and Mary Hesse's *Revolutions and Reconstructions in the Philosophy of Science* (Harvester, Brighton, 1980).

The literature on probability tends to be rather technical. Rudolf Carnap's *Logical Foundations of Probability* (Routledge & Kegan Paul, 1950), J. M. Keynes's *A Treatise on Probability* (Macmillan, London, 1921), H. Jeffreys's *The Theory of Probability* (Oxford University Press, 1939), L. J. Savage's *The Foundations of Statistics* (Wiley, New York, 1954), and Richard von Mises's *Probability, Statistics and Truth* (Hodge, London, 1939), are all standard works, as is Popper's *Logic on Scientific Discovery*. Donald Gillies expounds a version of the propensity theory in *An Objective Theory of Probability* (Methuen, London, 1973). Hugh Mellor in *The Matter of Chance* (Cambridge University Press, 1971) is philosophically interesting, as is A. J. Ayer's *Probability and Evidence* (Macmillan, London, 1972).

With reductionism, one of course begins to encroach on other areas of philosophy, such as the philosophy of mind. Popper's essay, referred to in the text (see Chapter 7 n. 1), in *The Open Universe* (Hutchinson, London, 1982) might be a useful starting-point, as would the book he co-authored with Sir John Eccles, *The Self and Its Brain* (Springer Verlag, Berlin, 1977). Functionalism regarding the mental is defended in various essays in Hilary Putnam's *Mind, Language and Reality* (Cambridge University Press, 1979), while eliminative materialism is defended by Paul Churchland, in *Scientific Realism and the Plasticity of Mind* (Cambridge University Press, 1979), and by Patricia Smith Churchland in *Neurophilosophy* (Bradford Books, Cambridge, Mass., 1986).

Versions of the 'strong programme' in the sociology of science are to be found in B. Barnes's *Scientific Knowledge and Sociological Theory* (Routledge & Kegan Paul, London, 1974), D. Bloor's *Knowledge and Social Imagery* (Routledge & Kegan Paul, 1976), and, to an extent, in Jonathan Powers's *Philosophy and the New Physics* (Methuen, London, 1985).

On the relation between science and value, Nicholas Maxwell's *From Knowledge to Wisdom* (Basil Blackwell, Oxford, 1984), as well as being interesting in itself, is an invaluable source of works similar in spirit. The most stimulating treatment I know of the relationship between the absolute conception of the world of science and other, more anthropocentric, perspectives is Thomas Nagel's *The View from Nowhere* (Oxford University Press, 1986), a topic also taken up in my own *The Element of Fire* (Routledge, 1988).

Index